A Tribute to Prof. Dr. Da Ruan

A Tribute to Prof. Dr. Da Ruan

Jie Lu and Etienne Kerre (Eds.)

A Tribute to Prof. Dr. Da Ruan

Springer

Editors

Jie Lu
University of Technology Sydney
Australia

Etienne Kerre
Ghent University
Belgium

ISBN 978-3-642-30306-7 e-ISBN 978-3-642-30307-4
DOI 10.1007/978-3-642-30307-4
Springer Heidelberg New York Dordrecht London

Library of Congress Control Number: 2012938135

Printed on acid-free paper

Springer is part of Springer Science+Business Media (www.springer.com)

This book is a present to Da who did not know he still had so many sweet friends who thought well of and miss him.

This book is dedicated to Da's family.

--Jie and Etienne

Foreword

Da Ruan's sudden passing away on July 31, 2011, at the age of 50, was a terrible shock to all who knew him. At the age of 50, Da Ruan was at the peak of his distinguished career. The author or co-author of over 270 papers and 36 books, Da Ruan was a highly visible figure on the international scene as a researcher, educator, an organizer of conferences, editor, lecturer, guest professor and, very importantly, as a bridge between Asian countries, especially China, and the West. But Da Ruan was much more than a prominent member of fuzzy logic and computational intelligence communities. He was a man of vision, initiative, great personal warmth and unbounded energy. His passing away is an irreparable loss—a loss which will be keenly felt in coming years.

In 1985, Da Ruan was one of two students selected by the Chinese government to study in Belgium toward a Ph.D. degree. He joined Professor Etienne Kerre's group on fuzziness and uncertainty modeling, and received his Ph.D. degree in 1990. After receiving his Ph.D. degree, Da Ruan became a postdoc at the Center for Nuclear Research in Mol. This was the beginning of his illustrious career and his lifelong association with Etienne. This volume, "A Tribute to Professor Doctor Da Ruan," edited by Jie Lu and Etienne Kerre is a collection of eloquent tributes to Da Ruan's life and work.

In most of my forewords, I spotlight a development in fuzzy logic and its applications which I deem to be of importance. In the instance of this foreword, I should like to comment briefly on what I consider to be an important trend—a trend toward a fusion of probability theory and fuzzy logic. Underlying this trend is the realization that, in large measure, uncertainty involves a mixture of fuzziness and probability. In this mixture, probabilities and events are commonly described in a natural language. I have long advocated the view that standard probability theory cannot deal with problems in which probabilities and events are described in a natural language. To deal with such problems, probability theory must be generalized through addition of concepts and techniques drawn from fuzzy logic and, more particularly, the formalism of Computing With Words (CWW).

An important concept which is the child of the marriage between probability theory and fuzzy logic is what I call a Z-number. I introduced this concept in my lecture at the World Conference on Soft Computing in San Francisco, in May 2011. After my lecture, I was approached by Da Ruan. His feeling was that the concept of a Z-number has the potential for many practical applications, and that he intended to explore what he felt was a new direction. It is a source of deep regret for me that his intention did not materialize.

Briefly, a Z-number is an ordered pair of fuzzy numbers, (A, B), in which A is a fuzzy value of a variable, X, and B is the fuzzy probability that X is A. Typically, A and B are described in a natural language. As a simple illustration, (usually, small) is a Z-number. An interesting question is: What is the square root of (usually, small)? Problems of this kind fall within the province of computation with Z-numbers.

The importance of the concept of a Z-number derives from the fact that in many realistic settings our knowledge of the value of a variable is uncertain. Computation with Z-numbers opens the door to construction of models in which the values of variables are certainty-qualified, e.g., usually, very likely, surely, etc. Clearly, in most realistic settings such models would be a better approximation to reality than models in which the values of variables are not qualified. An intriguing application of the concept of a Z-number is an if-then rule in which the consequent is usuality-qualified. A simple example is: If unemployment increases then usually stock market declines.

What would Da Ruan do with the concept of a Z-number? It is a question that can be raised but not answered. It is a painful realization that Da Ruan is not with us. He was a truly unique individual.

Lotfi A. Zadeh
Berkeley, California, USA

Prof. Da Ruan and Prof. Lotfi A. Zadeh in FLINS2008, Madrid

Preface

On 1 August, 2011, very sad news reached our scientific research community, we didn't believe it and we don't want to believe it, but it is true: *'Our dear friend, Prof. Da Ruan, suddenly passed away yesterday'*.

Only 50 years old, Da Ruan was in the prime of life, as a scientist, in the prime of career; no chance to prepare, and no time to say good-bye, suddenly left us for his heavenly abode, far away from his family and friends.

Our hearts were broken, we were full of sorrow, and tears streamed down our faces. How desperately we hoped that it was not true! The whole fuzzy logic research community was shocked and saddened by the loss of Da Ruan. At first, people questioned if it was indeed true. It seemed impossible – the loss of such a great man was truly shocking and caused heartfelt pain throughout the community. "Providence is always hard to ask for"—an incurable pain for us, for our hearts.

People found various ways to channel their grief for their close friend. Hundreds of emails quickly spread the tragic news throughout the world, in memory of Da. The large number of emails was testimonial to the high respect in which he was held, and how widely he was liked, both personally and professionally.

Some emails recalled fortunate meetings with Da and how he impressed people with his sunny smile and friendly personality. Others talked of their long-time research collaboration with Da. Some expressed appreciation to Da for helping to build their careers and achieve their goals. Some reflected on their first visit to Da at SCK•CEN and how he took care of them. There were expressions of appreciation for assistance with difficulties, and panegyrics about Da's brilliant speeches. All emails were full of accolades, admiration, high praise and honour for Da Ruan.

Indeed, this huge outpouring of grief, along with the email accolades and expressions of admiration, high praise and honour for this man, was the motivation and driving force behind this book. Producing and editing this book – to the memory of Da, dedicated to his family, for his colleagues, friends and students – has been a great honour and we hope that it will inspire future generations in the fuzzy logic community. We hope that it reflects Da's noble-minded and distinguished values and that it will have long-term significance in the research community.

Da Ruan was born in Shanghai where he spent a happy youth in the international metropolis. Full of vigour and dynamism, he completed his studies at Changzhi High School in 1979 with distinctive results, which earned him admission to Fudan University, a top Chinese university. After four years, he graduated with Honours and became an Associate Lecturer in the Department of Mathematics at Fudan University. In 1986, he was selected as an 'excellent youth scholar' to study at Ghent University, and in 1990, his PhD degree was conferred there. Since 1991, Da worked at SCK•CEN, starting as a post-doctoral researcher and progressing to a leading scientist position.

Da Ruan's short life was colourful and creative. He was a motivated and very productive scientist, publishing 36 books and 270 papers in journals and conferences. He led several very important fuzzy logic research projects in nuclear science, decision systems and risk analysis. His extraordinary contribution made him an undisputed leader and figure of authority in his fields of research.

He was a quick thinker and surpassed many others with his energy and indefatigability. He was an excellent leader. In 1994, at only 34 years, he took the initiative to establish the FLINS (Fuzzy Logic and Intelligent technologies in Nuclear Science) conference series, which grew to become the world's leading conference in the fuzzy information processing field. He continued to play a leading role, establishing the ISKE (Intelligent Systems and Knowledge Engineering) international conference series in 2006 – this has become a very popular conference in the area of intelligent systems. He also founded the *International Journal of Computational Intelligence Systems* (IJCIS), which, in the few years since its introduction, has acquired a very good reputation and international standing. He made outstanding contributions in the research fields of fuzzy logic, fuzzy decision making, computational intelligence, risk management and, in particular, the application of these advanced techniques in real-world nuclear science and nuclear industry. He was a very influential man throughout the world.

Da was very committed to his work and career, and was adept at handling complicated and challenging issues. He was confident, honest and open, his work was thorough, and pressure always excited him. His success can be attributed to a combination of his brilliance, expert knowledge of his scientific research, and his willingness to work hard and always push himself just that little bit further. He was therefore great fun to work with and people loved to contribute to the books he edited and attend the conferences he organised. As the undisputed leader of his community, Da's absence at the recent ISKE2011 conference held in Shanghai was consequently intensely felt.

Da was a charming and personable man with a cheerful and friendly personality and scholarly bearing. He had a playful sense of humour and liked to tell jokes and funny stories. He enjoyed fine food, particularly seafood, and he delighted in cooking delicious Chinese meals.

He loved his hometown, Shanghai, and his mother country, China. Over the last 20 years, he was a Visiting Professor at many universities in China and he supervised, trained and helped many Chinese scholars to become world-class researchers. He promoted effective communication between Chinese and Western cultures in our research community, benefiting both Chinese and Western scholars.

Da was a warm-hearted person who cared deeply about people and those he met all wanted to be his friend. He had an extensive community of friends and colleagues, and was very well-connected. People sought Da out for collaboration, so that he had over 100 published papers with researchers from other countries such as Australia, China, France, Italy, Spain, Turkey, UK and the US, all of whom are all proud of their collaborative experience with Da.

The authors of this book are all Da's friends; 88 academics from 47 institutes in 17 countries. They are respected senior, and junior, scholars; they are also Da's colleagues, classmates and peers; they are world-class researchers, and distinctive academics. The "father of fuzzy theory", Professor Lotfi Zadeh, 90 years old, has written the foreword of this book. Professor Yager, the Editor-in-Chief of the international journal *Intelligent Systems*, Professor Wang, ex-Editor-in-Chief and Professor Pedrycz, current Editor-in-Chief of the international journal *Information Sciences*, Professor Kacprzyk, Editor-in-Chief of *Studies in Computational Intelligence* by Springer, and Professor Zimmermann, the founder of the international journal *Fuzzy Sets and Systems*, all write of their particular experience with Da.

The majority of these authors knew Da well, becoming his friend through FLINS conferences. Although their expertise is in programming and mathematics, they have written touching stories of their experience with Da, expressing their grief at losing their dear friend. Reading these stories, we were deeply touched – and surprised – to find out that our scientists have become great writers.

Editing this book has not been easy. Emotions were constantly stirred throughout the process. Photos of Da are constant reminders of his loss; stories about him, his life, his work, re-awaken the grief. We have encouraged and supported each other throughout the editing process, and the book is a sincere tribute by all authors to Da Ruan.

Da's friends and colleagues have contributed over 60 chapters and more than 70 pictures, and we have been deeply moved by the articles. This has also been a gift to us, enabling us to better understand Da's contributions to the research community and come to know more about his life and his work. We more clearly appreciate his humanity and love of life, and his zeal, commitment and ambition in his academic life. Most importantly, editing this book about Da has taught us that, although knowledge and talent are important, it is sincerity, honesty and kindness that are the true indicators of the greatness of a person.

We wish to thank our all wonderful authors and the cover page designer, Mr. Chen. Our grateful thanks go to Tom and other Springer staff for their support in publishing this book.

Da has gone, but we are lucky to have been his friend. We are lucky to have worked with him. It is up to us to continue his work by providing a friendly, compassionate community and fun environment within a vibrant, exciting and motivated research community – to live up to Da's memory.

Da will remain in our hearts forever.

15 March 2012

Jie Lu
University of Technology Sydney
Australia

Etienne Kerre
Ghent University
Belgium

编者的话

噩耗惊曝，我们不敢相信，也不愿意相信，但它的确是真的。2011 年 8 月 1 日，如“六百里加急”，迅速把悲痛的消息传遍我们的学术同行——我们最心仪的朋友阮达昨日逝世。

年仅 50 岁，作为一个科学家，他风华正茂，春秋正富。可他就这么来不及思考，来不及准备，来不及期待，来不及告别，瞬间凋零，与朋友们从此阴阳永隔。

我们的心碎了，脑海里一片空白，懵懂之后，眼泪簌簌地往下淌。还一再地自问：“这是真的吗？” 也有好几位朋友在电话中痴痴质问：“他怎么就死了呢？” 那呜咽声，那嘶哑地呼唤，令人撕心裂肺。“天意从来高难问”，这将是我们心中永远的痛。

大家快速用各种方式表达自己失去挚友的悲痛。成百个电子邮件把这巨雷一样的消息传遍世界。邮件里，有的回忆如何幸运地结识阮达并成为他的朋友；有的温馨叙述与阮达心情畅适的合作情景；有的表达与阮达通信问业、多有借重的感激；有的追忆阮达给予的“扫榻以待”的礼遇；有的念及困难时分阮达如何施以援手；还有的是对阮达精彩演讲的由衷追捧。对于阮达“推诚相与”之悦，“大恩难报，大情难卸”之念，倾予了最响亮的颂扬与赞美。众人的感觉与感受竟是如此地契合！这一切，正是我们编辑这本书的强大动力和源泉。为了阮达，为了他的至尊和家人，为了他的朋友和学生，为了年轻的一代和下一代，留下这么一本书，不是赋有十分长远的意义和十分珍贵的价值么！

阮达 1960 年 9 月出生于中国上海，在这个智慧的城市度过了他难忘的青少年时光。 1979 年他以优异的成绩入读复旦大学这个中国顶尖学府，四年后毕业留校任教，1986 年作为优秀青年学者被推荐到比利时根特大学深造，1990 年获得博士学位。1991 年直至去世，一直在比利时核能研究中心工作，从博士后到高级研究员。

阮达短暂的一生是丰富而多彩的，贡献非凡。首先在专业上他是强者，一个智力和感受力极高的人，是一个多产的科学家。他出版了 36 本书，发表了 270 多篇文章，领导完成了多项重要研究项目。可称是业内学术研究的一位巨擘。

阮达才思敏捷，精力过人，还有着令人折服的领导才干。1994 年，年仅 34 岁，就领风气之先，创建了模糊逻辑和智能技术在核科学（Fuzzy Logic and Intelligent technologies in Nuclear Science, FLINS）国际会议系列，现已成为最具影响力的模糊信息处理国际会议之一。他继续发挥开山作用，又于 2006 年创建了智能系统和知识工程（Intelligent Systems and Knowledge Engineering， ISKE）国际会议系列，现在已成为一个非常受欢迎的智能系统国际会议。他还创建了《计算智能系统国际杂志》，几年时间便竖起了可观可感的信誉和地位。他在模糊逻辑、模糊决策、计算智能，特别是这些先进技术在核工业的应用，有着卓越的贡献。使他的影响跨越了时空与国界。

阮达非常擅长对纷繁事物进行梳理，有敏捷的应变能力，对承担的任何工作，都全力投入，一丝不苟，凡事都要做到最优；在高强度的工作中，充满睿智、亢奋和自信。所以，人们都喜欢和他一起工作，喜欢在他编著的书刊投稿，喜欢参加他组织的会议。以致最近参加 ISKE2011 会议，因为没有了阮达这个领军人物而颇感缺失什么。阮达之所以这等得到业界众人的拥戴，还与他的勤奋相关，为了追求自己生存的意义，他一直在跟自己赛跑。

阮达也是一个很有魅力的人。最突显的是他的阳光性格和骨子里的儒雅。他爱说一些令人捧腹的话，有搞笑的天分；对美食特别是海鲜，情有独钟；对故乡上海及上海文化，偏爱有加。

他爱上海，爱他的母国中国。多年来，他在多所中国高校兼职，培养了一批国际化学者，更促进了中西文化融通交流，使诸多中西方学者受益。在他的内心深处，处处充溢大情怀，时时彰显着他对母国的责任，他报国有成。

阮达的工作干得这么出色，更为重要的一点是他拥有一颗爱人惜友的心。对他人总有一种人性关怀，袍泽之情。往往共同打拼不求名份，是个极好合作的伙伴。自然广结善缘，有广泛的人脉，为大家架起了友谊的彩桥，更诠释给我们什么叫友谊。在阮达发表的文章和著作中，有 **100** 余篇是与他国学者合作完成的，这些来自于中国、英国、澳大利亚、土耳其、法国、西班牙、意大利、美国的合作者，无不为他们有与阮达合作的经历而自豪。一个人，在才华、人品、勤奋、能力诸方面，拥有一二，或许不难，难得的是像阮达，样样不缺。

这本书的作者都是阮达生前的好友。达 **80** 多人，来自 17 个国家。既有长者，也有后学；既有同仁，也有同门；既有学问顶级的前辈，也有资深大牌的业界巨子与学术界名流。“模糊数学之父”美国的扎德（Zadeh）教授，以 90 岁的高龄写下这本书的序。国际杂志《智能信息系统》主编雅各（Yager）教授，国际杂志《信息科学》原主编王（Wang）教授，现任主编裴椎兹（Pedrycz）教授，《Springer 计算智能研究丛书》主编、波兰院士卡波则克（Kacpkzyk）教授，国际杂志《模糊集与系统》创办主编齐默尔曼（Zimmermann）都写下他们和阮达相聚的独特感受和十分感人的情节。

这些作者中更多的是在 FLINS 会议上与阮达相识并和他成为朋友的人。从参加第一届 FLINS 的老朋友到 2011 年加入这个“阮达团队”的新同行，这些作者们用他们编程序的手，写公式的笔，来绘制他们和阮达共处的日子，表达自己的心曲与哀思。读着这些故事，我们的感受倍加，而且非常吃惊地发现我们这些科学家都已经变成了作家。

编辑这本书对我们来说真不是一件易事。这并非着眼它的工作量，而是一直承受着感情上的冲击。每一次看到阮达的照片，都心头一紧；每次读起阮达的故事，泪水都要在眼框里打转；特别当整理他的个人简历，那么辉煌的成就，可这么年轻就走了，阵阵痛惜，以至让人喘不过气来。这种万千思绪自始至终贯穿着成书的全过程。我们相互鼓励，相互支持，决心尽我们最大的努力编好这本书。这本圣洁又珍贵的书，可是全体作者送给阮达的一个无比真诚又无比温馨的供礼啊！

当我们完成了五十多个章节，七十多幅照片的编辑工作之后，我们深深地被我们的作者朋友感动了。作为阮达多年的友人，从这些章节里，我们看到了比我们知道得更多的故事，了解到了他更多、更高、更感人的大行。是阮达告诉我们何为大爱，还告诉我们：诚实理性的学术品格固然可尊，而真诚和善良比起才华更可敬；人格的力量时常比学识还要伟大！

阮达兄弟与世长辞。

阮达太累了。

我们为有这样一个朋友感到骄傲，为曾和他一起工作过感到幸福。

为今之计，我们只能承担起阮达对时代的许诺，接过他的大智和大爱，抽青吐翠。

阮达的精神丰碑，在我们的心中永存！

路节，悉尼科技大学，澳大利亚

Da in Sydney, Australia, Dec. 2005

Jie, Etienne and Da in Ghent University, Belgium, Jan. 2005

Da and his FLINS friends in Toledo, Spain, Sept. 2008

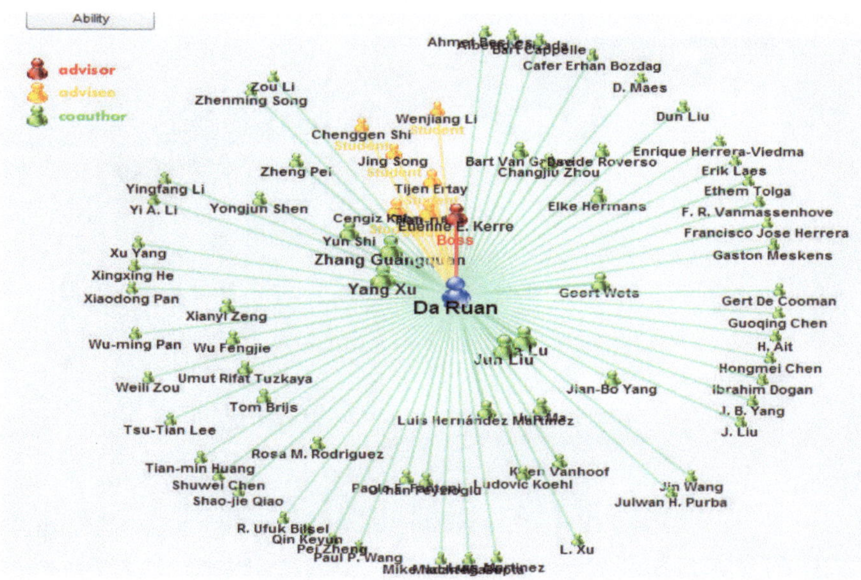

Da's research collaborations

Da chairing FLINS08, Madrid, Spain, Sept. 2008

Da chairing Soft Computing Panel in World Conference on Soft Computing, San Francisco, USA, May 25 2011

Jie, Guangquan and Da in IEEE Conf. on Systems, Man, and Cybernetics, San Antonio, USA

Da, Etienne and Guoqing at the China-Flanders workshop in Hasselt, Belgium, 2004

Da in his 50 year birthday party, Shanghai, China, 2010

Da in Australian Parliament House, Canberra, Australia, Nov. 2006

Contents

Part II: Da's Scientific Life

Part III: Da's Academic Activities

Part IV: Thoughts by Da Ruan's Colleagues and Friends

Part I
The Ceremony for Da

1.1 A Glimpse of the Flood of Emails after the Sad News Was Spread

"I would like to express my sympathy to his wife and family and tell them that I share their grief. I have met Da twenty years ago and I have always been impressed by his humanity, his empathy with his community, in addition to his great scientific achievements. I will always remember the last time I have seen him, last year in Chengdu, with his wife and sons, in so friendly moments. I will miss him a lot."

--- Professor Bernadette Bouchon-Meunier

"Please accept my heartfelt condolences on the unbelievable death of Da Ruan, also in the name of my colleagues at Springer in Heidelberg. Working and communicating with him always has been a great pleasure and honour for me and led to a large number of successful publications, combined with his outstanding friendliness and charming attitude. He was a wonderful person- we will very very much miss him!"

--- Professor Thomas Ditzinger

"I am writing this letter to express deep condolences on the untimely pass-away of Da Ruan. He had always been an excellent scientist to work with. He was a very true person with a golden heart and has always been there for support in different aspect. Da will be sorely missed. We know that there cannot be any amendment to the loss which has been incurred by Lord. Let this letter bring some relief in the painful times. God Bless you all."

--- Professor Hamido Fujita

"I am so sorry! I have known him for so many years and have always enjoyed all professional and personal contacts with him. I have also known his widow for whom I feel so sorry in particular because their children are still so young. He has been one of the most visionary people in our field who have always known how to combine theory and practice. We will miss him very much!"

--- Professor Janusz Kacprzyk

We lost a man of kindness, helpfulness and excellent research. He never forgot anyone anywhere; every friend was always in his mind. We will miss you, Da Ruan!"

--- Professor Cengiz Kahraman

"Dear Da, on behalf of all my colleagues I can tell you that saying goodbye to you is very hard. We will miss you so much: your true friendship, your optimism, your unselfish devoting with respect to all your commitments. Last Tuesday I went to say goodbye to your body and I saw your face expressing pride as it was saying: "I did my very best, please continue my work and be friendly to each other". Thank you, Da, for giving us so much love and friendship.

--- Professor Etienne Kerre

"I learned a lot with you, my dear Da, but I never forget is your smile, your winks, your friendly advice and your kind words."

--- Professor Victoria Lopez

"Well established researcher and internationally highly recognised expert in computational intelligence and decision making, Da Ruan launched in 1994 the FLINS conferences, successfully held since then. I will miss Da Ruan as one of my very best friends, a wonderful person and a hard worker, always kind and helpful."

--- Professor Javier Montero

"It is very difficult to find the right words to express all my feelings in this very sad moment. I will bring with me forever the memory of Da, his kindness, his openness, his availability, his correctness and his valuable scientific contributions."

--- Professor Gabriella Pasi

"This is terrible news. I still remember talking and laughing with Da in San Francisco in May. Da - you will be missed a lot!"

--- Professor Marek Reformat:

"The passing away of our friend Da is indeed very sad news for all of us who knew him and had the pleasure of working closely with him. My thoughts go to his family and his nearest friends."

--- Professor Davide Roverso

"I still cannot believe that it is true. Da was so energetic, full of brilliant ideas, and enthusiastic about his research, teaching and collaboration. He was emailing us just a few weeks ago. I can still picture Da's vivid presentation, and hear his energetic voice in discussions and debate. It is a huge loss to all of us."

--- Professor Dong-Ling Xu

"It has been a so shocked and terrible news about Da! I cannot believe it was true. I still can remember clearly the first time we met each other in a conference in USA, from that moment, we have been very good friends for many years! He is a so nice person and so excellent scientist! He only became 50 years, a terrible loss to us all! Da will be missed by everyone who knew him. Our thoughts and prayers will go with him. My sincere and solemn condolences to Da' family."

--- Professor Yang Xu

"I was very saddened to hear of the death of Da Ruan. I warmly remember the meal on the beach in Sydney with Da, Da enjoyed himself very much on that occasion. I am glad I had the opportunity to see Da at the conference in San Francisco...."

--- Professor Ronald Yager

"No word can express our shock at the sad news. I still could not believe that Da will no longer be with us since I was told the terrible news last night. At a meeting at York University today, Dong-Ling and I shared this sad news with colleagues with shock and disbelief. Da is our dear friend and a most prolific and respected researcher worldwide. We just discussed issues via email a few weeks ago when I was in Australia attending the 2011 IFORS conference. He already accepted our invitation to visit Manchester later this year. His sudden departure is a huge loss to our research community in general and to our long term collaboration between SCK and Manchester in particular. He will be remembered by us forever. Our deepest condolence goes to Greet and the two children. At this most difficult time, we all share the terrible loss of Da.

--- Professor Jian-bo Yang

"I was deeply shocked and saddened by the news that Da Ruan suddenly passed away. It is an irreparable loss for the fuzzy logic community world wide."

--- Professor Lotfi Zadeh

"It is really a very very sad news. I can not believe my eyes when reading these words. I highly appreciate Da for his intelligence and kindness to all his friends. We even discussed on our future collaboration three weeks ago before my trip to China. We need to do something together for him. We will remember Da forever..."

--- Professor Xianyi Zeng

1.2 Chinese Memorial Plaque and Memoriam Book

A Chinese memorial plaque by Da's Chinese friends

A commemoration book for Da, made by Guangquan and Jie

1.3 Summary of the Ceremony for Da Ruan at Vito Conference Hall, Mol, Belgium, 6 Aug. 2011

Jun Liu

University of Ulster, Northern Ireland, UK

2011年8 月 6 日 11 时，阮达教授的葬礼在他曾经工作地： Mol, SCK/VITO conference hall 隆重举行。前来参加的有阮达生前的领导，同事，教授，远近朋友，家属及从上海专程赶来的妹妹及外甥女，估计有超过500人。葬礼旨在纪念和庆祝阮达成功和辉煌的一生. 仪式由阮达全家的好友MIA女士主持. 整个仪式组织的非常专业和正式，音乐和歌声伴随着特别来宾的悼念致辞，使所有来宾在对达的失去无限悲伤而哭泣的同时，也充满了神圣的氛围，又给来宾带回对达的深情的回忆和思念，以及安宁和安慰! 所有的致辞者和表演者都是达的非常亲密的亲人，领导，导师和好友. 大荧幕上展现的是阮达面带微笑，神采奕奕的近照，让来宾们再一次不能相信他真的已经离我们而去。阮达的骨灰盒被悼念他的花圈花篮环绕着，有中国驻比利时使馆，SCK, 旅比华人专业人士协会，以及一些大学，生前亲人和中外朋友同事等送献的。葬礼仪式持续 大约两小时，在所有来宾对达的全家的安慰和祝福中结束.

The memorial service ceremony for Prof. Da Ruan took place on Saturday, August 6 at 11 a.m. in the SCK/VITO Conference room, Boeretang 200, 2400 Mol, Belgium, where Da lived and worked. The ceremony aimed and focused on remembering and celebrating the wonderful life of Da. Over 500 people including Da's colleagues in SCK and universities, friends and family relatives from many cities of Belgium as well as Australia, UK, Spain, Germany, France, Mexico, and other countries attended the ceremony especially Da's sister and niece from Shanghai.

The Master of ceremonies was Mia the closest friend of Da's family. The ceremony was organized in a very formal and professional way, followed by recorded music, the singing of hymns, and individual speeches including Da’s sister Fei Ruan, the General Director of SCK Prof. Eric van Walle, Da friends Prof. Etienne Kerre and Prof. Jie Lu. There was deep grief, but more of a focus on memory, peace and comfort.

All speakers and musicians on stage had a very close personal relationship with Da. The big screen on stage had a slide show of Da's latest pictures full of smiles and energy, which again and again make us recall the fragments of life which will stay so close in our memory. Da's funeral urn was surrounded by floral tributes sent by the Chinese Embassy in Belgium, SCK, family, some universities and institutes, and close friends. The ceremony lasted for about two hours and finished with all the guests meeting with Da's family members in person and giving their sympathy and condolence to them.

1.4 Speech in the Ceremony for Da Ruan at Vito Conference Hall, Mol, Belgium, 6 Aug. 2011

Etienne Kerre

Ghent University

Beste Greet, Alexander, Nicolas en Christiane
Dear Fei and Di,
Dear friends and colleagues,

Wat een vreselijke tijd! Wie had een week geleden kunnen vermoeden dat wij vandaag afscheid zouden moeten nemen van Da? Het nieuws sloeg bij ons in als een bom. Da, ogenschijnlijk gezond van lijf en leden, is niet meer. Hoeveel e-mails en telefoons van ongeloof en verbijstering hebben we de voorbije dagen niet met elkaar gewisseld. En de eeuwige vraag: waarom,waarom hij? Zo een in-goed en getalenteerd man! Zo een liefhebbende echtgenoot en vader! Zo een vriend van onschatbare waarde! And such a loss for our fuzzy community!

I met Da for the first time about 25 years ago when he joined my research group on Fuzziness and Uncertainty Modelling at Ghent university. That time every year China selected two excellent master students to go abroad in the framework of a bilateral cooperation between China and Belgium to work towards a PhD degree. From the very beginning of our cooperation I was very touched by his competence and hard working to tackle completely new issues in the recently developed mathematics of fuzziness. Da obtained his PhD in mathematics from Ghent University in 1990. After obtaining his PhD he could get a post-doc position at the Center for Nuclear Research in Mol, where he started his very successful career as a researcher combining theoretical as well as applied contributions and especially as the project leader of his scientific child called FLINS (an acronym for Fuzzy Logic and Intelligent technologies in Nuclear Science). Indeed unbelievable successful!

Da has authored about 120 peer-reviewed papers in leading international journals, five textbooks, about 100 book chapters and conference papers on different soft computing issues in decision support systems, aggregation operators, information management and nuclear systems, to call only a view of them. Indeed as Prof. Janusz Kacprzyk from the Polish Academy of Science wrote: “Da has been one of the most visionary people in our field who has always known how to combine theory and practice”.

Da has edited 20 research books for world publishers such as Springer and Kluwer and 17 special issues on artificial intelligence for top journals such as Fuzzy Sets and Systems, Information Sciences and International Journal of Intelligent Systems.

But as said before his major scientific achievement was the founding of FLINS with its international conference held every two years starting from 1994, first in Mol and later on in Ghent, Antwerp, Blankenberge, Bruges, Genova, Madrid and Chengdu. Da was the main editor of the proceedings, not only responsible for its contents but also for the lay-out, including the design and colours of the cover. Da was an aesthete: everything is important, not only the contents but also the presentation.

Nobody can imagine that the planned 2012 FLINS conference in Istanbul will not be opened by Da.

Two months ago we attended the World Conference on Soft Computing in San Francisco to celebrate the 90th anniversary of the founder of fuzzy logic Prof Lotfi Zadeh. At that conference Da organized a panel discussion on "Soft computing challenges and future directions in society and industry". Da could convince top researchers from Europe, USA, Australia, Mexico to take part in this discussion on the future of our research. Afterwards I told Da that I was very proud to see him chairing this event in front of a completely full Jack Adams hall of the San Francisco State University.

Because of his highly rated scientific achievements and management capacities Da passed very strong criteria to obtain a guest professorship at Ghent University in 2005. In 2007 he also became a guest professor at Hasselt University, as well as adjunct professor in the University of Technology in Sydney and different universities in Shanghai, where he was praised for his lectures on fuzzy logic and its applications. Despite his so young age he obtained already an honorary doctorate from the Nuclear Power Institute of China in Chengdu.

Dear friends, nobody can imagine the energy Da has put into these achievements. Quoting Lotfi Zadeh, the founder of the fuzzy set theory: "Da Ruan was a leading member of the fuzzy logic community world-wide. His passing away is an irreparable loss…"

Maar Da heeft niet enkel intellectuele en wetenschappelijke verdiensten. Wat Da als mens betekende kan niet genoeg worden benadrukt. Wat de betekenis is van vriendschap heeft Da aan zo velen van ons duidelijk gemaakt. Getuige ook waarvan de tientallen e-mails die de voorbije dagen het heengaan van Da hebben betreurd. Ikzelf verlies aan Da één van mijn allerbeste vrienden.

Da was ook een fantastische echtgenoot, vader, zoon en broer.

Hoe fier was hij niet op jou Greet, en op je talenten: je zangtalent, je schrijftalent…

En dan zijn twee zonen, zijn oogappels, eveneens getalenteerd en op weg prachtige mensen te worden zoals hun vader. Hoe jammer dat hij niet meer zal kunnen genieten van jullie verdere successen!

And how full of love and loyalty he remained with respect to his parents, sister and brothers and to his beloved China. Dear Fei, we can feel your sorrow: my wife's brother passed away at the age of 48 in completely similar circumstances.

Everybody could experience the enthusiasm he showed in praising the Chinese culture to his foreign friends. We all know his passion for the good Chinese food. He was an excellent cook. I lively remember his talks with Ron Yager from New York about Chinese delicacies. You know Da that we intended to open after your retirement an exclusive restaurant with high- class Chinese food...

Dear Da, on behalf of all my colleagues I can tell you that saying goodbye to you is very hard. We will miss you so much: your true friendship, your optimism, your unselfish devoting with respect to all your commitments. Last Tuesday I went to say goodbye to your body and I saw your face expressing pride as it was saying: "I did my very best, please continue my work and be friendly to each other."

Thank you, Da, for giving us so much love and friendship.

E. Kerre received from Da the FLINS Golden Chair Medal in 2010 in Chengdu for his continuous support to FLINS conferences

A proof of a true friendship since 1987

1.5 Poem Recitation in the Ceremony for Da Ruan at Vito Conference Hall, Mol, Belgium, 6 Aug. 2011

Jie Lu

University of Technology, Sydney, Australia

致阮达

你是阳光，把温暖慷慨地抛向每个朋友；
你是巨树，以宽厚的胸怀揽着众人乘凉；
你是智星，以绝顶的聪慧点燃众友人的潜能，
大家肩并肩手牵手，浩浩荡荡上征程。
MOL，因为有你充满魅力而扬名四海，
随着你的离去，它又将变成友人们心中一块 抹不掉的忧伤地。
你好像不懂得拒绝，你好像不知道疲倦，
在你的生活里，一天不止24小时，一年不止365天。
达，你走的太匆忙，如大海波澜；
你留给我们的遗憾太多，如山雨云烟：
即将出炉的学说，尽力筹办的会议，尚未开口的精彩，即将完成的课题……
阮达兄弟，你太累了，请安息；
阮达兄弟，你走好，别太多惦记；
你无比的才学，无暇的人品，无尚的性格，留给我们无尽的回忆。

To Da Ruan

Your friendship, like the sun, nurtured us with your warmth.
Your leadership, like a big tree, provided us with strength, shelter and protection.
You like a star, gave us guiding light so we could achieve so much with our research.
You like a magnet, attracted and inspired all of us through FLINS.
You always did your best for your friends
You never knew how to take a rest yourself
In your life,
One day was more than 24 hours
One year was not only 365 days.
Mol became popular through you and
Now is a place that we will always cherish.
You left us too early
Your life is too short
With your passing we are left with a huge void: incomplete books, unfinished papers, draft proposals and conference keynote speeches which only stay on the web
Da, you will always be alive in our hearts;
Memories of you will stay with us forever.

Jie, Guangquan and Da's Friends

1.6 Speech in the Ceremony for Da Ruan at Vito Conference Hall, Mol, Belgium, 6 Aug. 2011

Hong Chen

VITO, MOL, Belgium

In the dormitory of SCK-CEN, Mol Belgium, 1993 (from right to left: Lian Wang, Hong Chen, Lili Wang, Da Ruan)

I am a friend of Da. Probably one of his oldest friends from Mol. I got to know him 20 years ago when he started at SCK. It all begins in the kitchen of SCK's dormitory. One day, came a group of very proud young engineers and PhDs. They are all from Ghent University. They labeled themselves as 'Ghent connections'. They drink 'Duvel' and talk about incomprehensive things such as fuzzy logic. Some of them even play bridge (you know whom I am talking about). Da was a very active member of the group. Soon the group extended to tenth of young people not only from Ghent but from other universities and also from different parts of the world. Back then, we are always together, going-out, partying, doing sports, falling in love, building families and having children…

All these years, the Da we know is a gentle, loving, and care friend. We never forget those days: playing cards all night long and Da always wins; eating Chinese food together and with no exception Da only likes those stinkdry fish and those look like thousand years old Chinese eggs; Da never turns out to be a big drinker so we used to ask him to drive home after the parties but then found out Da was then not really a good driver neither – still remember he crashed his first car so hopelessly on a tree. Yes, that is Da. In life, Da is not fuzzy, he is so real, distinctive, and passionate. Today, just like 20 years ago, Da again brings us together, but only with so much more added values. We will miss you, Da, but we will go on, only this time with so many nice memories about you and inspirations from you.

I would like to say a few words in Chinese, representing the Chinese community here: 安息吧，达，我们会永远怀念你！

1.7 Obituary: In Memoriam Prof. Dr. Da Ruan

Available online at www.sciencedirect.com
ScienceDirect
Fuzzy Sets and Systems 185 (2011) 1–4
FUZZY sets and systems
www.elsevier.com/locate/fss

Obituary: In Memoriam Prof. Dr. Da Ruan

Prof. Dr. Da Ruan

On July 30, 2011 our colleague and friend Dr. Da Ruan suddenly passed away in his home at the age of 50 due to a heart failure. I still cannot believe I have to write this in memoriam for one of the best human beings I ever met.

I met Da for the first time about 25 years ago when he joined my research group on Fuzziness and Uncertainty Modelling at Ghent University. That time every year China selected two excellent master students to go abroad in the framework of a bilateral cooperation between China and Belgium to work towards a Ph.D. degree. From the very beginning of our cooperation I was very touched by his competence and hard working to tackle completely new issues in the recently developed mathematics of fuzziness. Da obtained his Ph.D. in mathematics from Ghent University in 1990. After obtaining his Ph.D. he could get a post doc position at the Center for Nuclear Research in Mol, where he started his very successful career as a researcher combining theoretical as well as applied contributions and especially as the project leader of his scientific child called FLINS (an acronym for Fuzzy Logic and Intelligent technologies in Nuclear Science). Indeed unbelievable success!

Da has authored about 120 peer-reviewed papers in leading international journals, five textbooks, about 100 book chapters and conference papers on different soft computing issues in decision support systems, aggregation operators, information management and nuclear systems, to call only a view of them. Indeed as Janusz Kacprzyk wrote in a recent email: "Da has been one of the most visionary people in our field who has always known how to combine theory and practice".

Da has edited 20 research books for world publishers such as Springer and Kluwer and 17 special issues on artifi- cial intelligence for top journals such as Fuzzy Sets and Systems, Information Sciences and International Journal of Intelligent Systems.

But as said before his major scientific achievement was the founding of FLINS with its international conference held every two years starting from 1994, first in Mol and later on in Gent, Antwerp, Blankenberge, Bruges, Genova, Madrid and Chengdu. Da was the main editor of the proceedings, not only responsible for its contents but also for the lay-out, including the design and colours of the cover. Da was an aesthete: everything is important, not only the contents but also the presentation.

It is hard to believe that the planned 2012 FLINS conference in Istanbul will not be opened by Da.

Because of his highly rated scientific achievements and management capacities Da passed very strong criteria to obtain a guest professorship at Ghent University in 2005. In 2007 he also became a guest professor at Hasselt University, as well as adjunct professor in the University of Technology in Sydney and different universities in Shanghai, where he was praised for his lectures on fuzzy logic and its applications. Despite his so young age he obtained already a honorary doctorate from the Nuclear Power Institute of China in Chengdu.

Dear friends, nobody can imagine the energy Da has put into these achievements. Quoting Lotfi Zadeh: "Da Ruan was a leading member of the fuzzy logic community worldwide. His passing away is an irreparable loss…"

Da was also a tender husband for his wife Greet and an excellent father for his two sons Alexander and Nicolas. He too stayed full of love and concern with respect to his Chinese relatives: his parents, his sister and brother.

Everybody could experience the enthusiasm he showed in praising the Chinese culture to his foreign friends. We all know his passion for the good Chinese food. He was an excellent cook. I lively remember his talks with Ron Yager about Chinese delicacies.

Dear Da, on behalf of all my colleagues I can tell you that saying goodbye to you is very hard. We will miss you so much: your true friendship, your optimism, your unselfish devoting with respect to all your commitments. I went to say goodbye to your body and I saw your face expressing pride as it was saying: "I did my very best, please continue my work and be friendly to each other."

Thank you Da, for giving us so much love and friendship.

Books and Edited Volumes by Da Ruan

1. Jin, X., Liu, Y., Li, T., Ruan, D. (eds.): Proceedings of the 2010 International Conference on Intelligent Systems and Knowledge Engineering. IEEE Press, Beijing (2010)
2. Ruan, D., Li, T., Xu, Y., Chen, G., Kerre, E.E.(eds.): Computational Intelligence: Foundations and Applications, the Proceedings of FLINS2010. World Scientific, Singapore (2010), `http://www.worldscibooks.com/compsci/7939.html`
3. Ruan, D. (ed.): Computational Intelligence in Complex Decision Systems. Atlantis Press, World Scientific, Paris, Singapore (2010), `http://www.worldscibooks.com/compsci/1015.html`
4. Pei, Z., Ruan, D., Liu, J., Xu, Y. (eds.): Linguistic Valued Based Intelligent Information Processing: Theory, Methods, and Applications. Atlantis Press, World Scientific, Paris, Singapore (2009), `http://www.worldscibooks.com/compsci/`
5. Vanhoof, K., Ruan, D., Li, T., Wets, G. (eds.): Intelligent Decision Making Systems, the Proceedings of ISKE 2009. World Scientific, Singapore (2009), `http://www.worldscibooks.com/compsci/7616.html`
6. Wang, X., Ruan, D., Kerre, E.E.: Mathematics of Fuzziness—Basic Issues. Springer, Heidelberg (2009), `http://www.springer.com/978-3-540-78310-7`
7. Li, S., Li, T., Ruan, D. (eds.): Proceedings of the 2008 International Conference on Intelligent Systems and Knowl- edge Engineering. IEEE Press, Beijing (2008)
8. Ruan, D., Montero, J., Lu, J., Martinez, L., D'hondt, P., Kerre, E.E.: Computational Intelligence in Decision and Control, the Proceedings of FLINS 2008. World Scientific, Singapore (2008), `http://www.worldscibooks.com/compsci/6802.html`
9. Ruan, D., Hardeman, F., van der Meer, K. (eds.): Intelligent Decision and Policy Making Support Systems. Springer, Heidelberg (2008), `http://www.springer.com/978-3-540-78306-0`
10. Wang, P.P., Ruan, D., Kerre, E.E. (eds.): Fuzzy Logic—A Spectrum of Theoretical & Practical Issues. Springer, Heidelberg (2007), `http://www.springer.com/engineering/book/978-3-540-71257-2`
11. Lu, J., Zhang, G., Ruan, D., Wu, F.: Multi-Objective Group Decision Making—Methods, Software and Applications with Fuzzy Set Technology. Imperial College Press, London (2007), `http://www.worldscibooks.com/business/`
12. Li, T., Xu, Y., Ruan, D. (eds.): Proceedings of the 2007 International Conference on Intelligent Systems and Knowledge Engineering. Atlantis Press, Paris (2007)
13. Zeng, X., Li, Y., Ruan, D., Koehl, L. (eds.): Computational Textile. Springer, Heidelberg (2007), `http://www.springer.com/engineering/book/978-3-540-70656-4`
14. Lu, J., Ruan, D., Zhang, G. (eds.): E-Service Intelligence. Springer, Heidelberg (2007), `http://springer.com/978-3-540-37015-4`
15. Ruan, D., D'hondt, P., Fantoni, P.F., De Cock, M., Nachtegael, M., Kerre, E.E. (eds.): Applied Artificial Intelligence. World Scientific, Singapore (2006), `http://www.worldscibooks.com/compsci/6049.html`

16. Ruan, D., Chen, G., Kerre, E.E., Wets, G. (eds.): Intelligent Data Mining. Springer, Heidelberg (2005), http://www.springer.com/engineering/book/978-3-540-26256-5
17. Kerre, E.E., Huang, C., Ruan, D.: Fuzzy Sets Theory and Approximate Reasoning. Wuhan University Press, Wuhan (2004) (in Chinese)
18. Ruan, D., D'hondt, P., De Cock, M., Nachtegael, M., Kerre, E.E. (eds.): Applied Computational Intelligence, the Proceedings of FLINS 2004. World Scientific, Singapore (2004), http://www.worldscibooks.com/compsci/5563.html
19. Ruan, D., Zeng, X. (eds.): Intelligent Sensory Evaluation. Springer, Heidelberg (2004), http://www.springer.com/engineering/production+eng/book/978-3-540-20324-7
20. Xu, Y., Ruan, D., Qin, K., Liu, J.: Lattice-Valued Logic—An Alternative Approach to Treat Fuzziness and Incomparability. Springer, Heidelberg (2003), http://www.springer.com/engineering/book/978-3-540-40175-9
21. Zhou, C., Maravall, D., Ruan, D. (eds.): Autonomous Robotic Systems—Soft Computing and Hard Computing Methodologies and Applications. Physica-Verlag, Heidelberg (2003), http://www.springer.com/computer/artificial/book/978-3-7908-1546-7
22. Ruan, D., Fantoni, P.F. (eds.): Power Plant Surveillance and Diagnostics—Applied Research with Artificial Intelligence. Springer, Heidelberg (2002), http://www.springer.com/engineering/power+engineering/book/978-3-540-43247-0
23. Ruan, D., D'hondt, P., Kerre, E.E. (eds.): Computational Intelligent Systems for Applied Research, the Proceedings of FLINS 2002. World Scientific, Singapore (2002), http://www.worldscibooks.com/compsci/5020.html
24. Ruan, D., Kacprzyk, J., Fedrizzi, M. (eds.): Soft Computing in Risk Evaluation and Management—Applications in Technology, Environment and Finance. Physica-Verlag, Heidelberg (2001), http://www.springer.com/business/finance/book/978-3-7908-1406-4
25. Ruan, D., Aït Abderrahim, H., D'hondt, P., Kerre, E.E. (eds.): Intelligent Technologies and Soft Computing in Nuclear Science and Engineering, the Proceedings of FLINS 2000. World Scientific, Singapore (2000)
26. Ruan, D., Kerre, E.E. (eds.): Fuzzy If–Then Rules in Computational Intelligence: Theory and Applications. Kluwer Academic Publishers, Boston (2000)
27. Ruan, D., Huang, C. (eds.): Fuzzy Sets and Fuzzy Information-Granulation Theory—Key Selected Papers by Lotfi Zadeh. Beijing Normal University Press, Beijing (2000) (both in English and in Chinese)
28. Ruan, D., Huang, C. (eds.): Fuzzy Sets and Operations Research for Decision Support—Key Selected Papers by Hans-Jürgen Zimmermann. Beijing Normal University Press, Beijing (2000)
29. Ruan, D. (ed.): Fuzzy Systems and Soft Computing in Nuclear Engineering. Physica-Verlag, Heidelberg (1999), http://www.springer.com/engineering/power+engineering/book/978-3-7908-1251-0
30. Ruan, D., Aït Abderrahim, H., D'hondt, P., Kerre, E.E. (eds.): Fuzzy Logic and Intelligent Technologies for Nuclear Science and Industry, the Proceedings of FLINS 1998. World Scientific, Singapore (1998)
31. Ruan, D. (ed.): Intelligent Hybrid Systems: Fuzzy Logic. Kluwer Academic Publishers, Boston (1997)

32. Ruan, D. (ed.): Fuzzy Logic Foundations and Industrial Applications. Kluwer Academic Publishers, Boston (1996), `http://www.springer.com/math/book/978-0-7923-9774-8`
33. Ruan, D., D'hondt, P., Govaerts, P., Kerre, E.E. (eds.): Intelligent Systems and Soft Computing for Nuclear Science and Industry, the Proceedings of FLINS 1996. World Scientific, Singapore (1996)
34. Ruan, D. (ed.): Fuzzy Set Theory and Advanced Mathematical Applications. Kluwer Academic, Boston (1995)
35. de Cooman, G., Ruan, D., Kerre, E.E. (eds.): Foundations and Applications of Possibility Theory, the Proceedings of FAPT 1995. World Scientific, Singapore (1995)
36. Ruan, D., D'hondt, P., Govaerts, P., Kerre, E.E. (eds.): Fuzzy Logic and Intelligent Technologies in Nuclear Science, the Proceedings of FLINS 1994. World Scientific, Singapore (1994)

Etienne E. Kerre
E-mail address: etienne.kerre@UGent.be

1.8 In Memoriam EUSFLAT Board Member Da Ruan: Obituary

Etienne Kerre, Luis Martínez, Jie Lu, and Javier Montero

Mathware & Soft Computing Magazine. Vol. 18 n. 1 15 / 71

IN MEMORIAM

EUSFLAT board member Da Ruan: Obituary

By Professor Dr. Etienne E. Kerre

Midsummer 2011. Da prepared an extensive delicious Chinese dinner for his family and some friends. Everybody enjoyed the pleasant evening. But suddenly Da felt unwell: a heart failure made a dramatic end to his life and left us behind with memories about this remarkable friend and colleague.

Da was born and grown up in Shanghai, China. In 1983 he obtained a Bachelor of Science degree in Computational Mathematics from Fudan University, one of the top 4 universities in China. Due to excellent study results he became an asistant of this university. In 1987 he got a grant from the Chinese government to study abroad. In that way Da became my first Chinese PhD student, an intelligent and hardworking young researcher, capable to tackle new problems in the nascent mathematics of fuzziness. Da successfully defended his PhD on "A critical study of widely used fuzzy implication operators and their influence on the inference rules in fuzzy expert systems" in December 1990 with Dr Han Zimmermann and late Dr Phillippe Smets among the members of the jury.

As PhD student, he was an intelligent and hardworking researcher, capable to tackle new problems in the nascent mathematics of fuzziness.

In 1991 till his death Da became a permanent staff member of the Belgian Nuclear Research Center in Mol, Belgium. Among the long list of activities and initiatives Da has taken in that research center I will only mention two of them: first the development of a time-dependent three-dimensional finite-element diffusion cone TRANSFUSION for nuclear oil well logging applications, and second the building of a fuzzy logic control demo model that has been extended for on-line experiments at the BR1 nuclear reactor in Mol. Due to these advanced and daring applications of fuzzy logic Da has been hired as principal investigator by the International Atomic Energy Agency in Vienna to develop a fuzzy logic based intelligent tool for uncertainty analysis for nuclear safeguards applications as well as to develop an adaptive fuzzy controller for the TRIGA Mark III reactor in Mexico.

Da was very concerned and full of love for his family in Belgium (his wife Greet and two sons Alexander and Nicolas) and in China (his parents, brother and sister)

Da has also been an excellent researcher with more than 100 papers in internationally high ranked journals such as Fuzzy Sets and Systems, Information Sciences, Journal of Global Optimization, Journal of Multiple Valued Logic and Soft Computing, Journal of Computational Intelligence Systems, Journal of Intelligent Systems, Journal of Intelligent Systems, Journal of Approximate reasoning, Journal of Uncertainty, Fuzziness and Knowledge-Based Systems. Soft Computing, Mathware and Soft Computing. His main topics of research were: fuzzy implications, decision support systems, risk evaluations, lattice-valued logics and reasoning with fuzzy if-then rules.

But Da will ever be remembered for his excellent service to our fuzzy community. First of all he edited close to 20 special issues for several international journals. Secondly he edited 36 books on computational intelligence, mathematics of fuzziness and intelligent decision making systems. He has been a referee for about 25 international journals. Moreover in 2007 he launched as editor-in-chief the International Journal of Intelligence Systems (with already an impact factor of 1.471 in 2010!) as well as the book series "Computational Intelligence Systems".

Da has been an excellent researcher with more than 100 papers in internationally high ranked journals. He will ever be remembered for his excellent service to our fuzzy community

However his major scientific achievement was the founding of FLINS, an acronym for Fuzzy Logic and Intelligent technologies in Nuclear Science. Every 2 years from 1994 on Da organized the international FLINS conference that could count on a continually growing number of participants. Da was always proud to have the prestigious proceedings of this conference ready at the start of the conference.

We will miss Da so much as an excellent friend, full of kindness and unconditional helpfulness

Because of his excellent CV Da became a guest professor at several universities in Belgium (Ghent and Hasselt), Australia (UTS in Sydney) and China (Shanghai Electricity and Power university, Shanghai Maritime university, Shanghai university, Xihua university in Chengdu, Southwest Jiaotong university in Chengdu): he was a beloved teacher, Ph D promoter and research guide. Looking at these achievements one can hardly imagine the amount of energy Da has put into his scientific activities. But Da was also very concerned and full of love for his family in Belgium (his wife Greet and two sons Alexander and Nicolas) and in China (his parents, brother and sister). Many of us have experienced his passion against well-prepared Chinese food.

We will miss Da so much, not only as a perfect organizer of meetings, not only as an indefatigable editor of special issues of journals and books, not only as an intelligent mind to develop high technological control devices... but most importantly as an excellent friend, full of kindness and unconditional helpfulness. Thank you Da!

By Professors Dr. Luis Martínez, Dr. Jie Lu and Dr. Javier Montero

Our colleague Da Ruan, member of the EUSFLAT board since the last EUSFLAT general assembly, unexpectedly passed away on 31 July, 2011 from heart failure at the early age of 50 years. His death was a shock to all of us. He will be very sadly missed. Da Ruan was a leading member of the global fuzzy logic community, an appreciated EUSFLAT member, and Editor in Chief of the EUSFLAT official journal, the International Journal of Computational Intelligence Systems.

Da Ruan's untimely passing has created much sadness in our research community and poses an irreparable loss to the fuzzy logic research field. He was very energetic, full of brilliant ideas, and always enthusiastic about his research, teaching and collaboration. A number of Web pages have been created around the world to honour Da Ruan's memory (see, e.g., http://decide.it.uts.edu.au/DaRuan/Da_Ruan.html, which contains many world-class researchers' messages in tribute to Da's scientific work and his excellent character; see also http://www.sckcen.be/fr/Nouvelles/In-Memoriam-Da-Ruan). He was well liked and highly regarded, as evidenced by the messages of tribute.

Da Ruan was a principal research scientist at the Belgian Nuclear Research Centre (SCKoCEN). He received a PhD in Mathematics from Ghent University, Belgium in 1990. His major research interests were in the areas of mathematical modelling, computational intelligence methods, uncertainty analysis, decision support systems and AI applications to information management, cost/benefit analysis, nuclear safety and security related fields. He published 90 peer-reviewed journal articles, two text books, and 20 research books on artificial intelligence, decision support systems, and intelligent information systems and applications.

Da Ruan was a leading member of the global fuzzy logic community, an appreciated EUSFLAT member and Editor in Chief of the EUSFLAT official journal

Da Ruan received an honorary doctoral degree from the Nuclear Power Institute of China for his research achievements, including intelligent control for nuclear reactors in 1995. He served as a member of the editorial board of Fuzzy Sets and Systems (Elsevier), as regional editor for the European International Journal of Intelligent Automation and Soft Computing (TSI Press, Albuquerque, NM), as co-editor-in-chief of the International Journal of Nuclear Knowledge Management (Interscience Publishers, Geneva), as editor-in-chief of the International Journal of Computational Intelligence Systems (Atlantis Press, Paris), and as editor of the book series on Intelligent Information Systems and for the proceedings series of Computer Engineering and Information Science (World Scientific,

Web pages to honour Da Ruan's memory

- http://decide.it.uts.edu.au/DaRuan/Da_Ruan.html
- http://www.sckcen.be/fr/Nouvelles/In-Memoriam-Da-Ruan

Singapore). He was a guest Professor at the Dept of Applied Math and Computer Science in Ghent University and at the Dept of Applied Economics in Hasselt University, respectively, in Belgium and an Adjunct Professor in the Faculty of Information Technology at University of Technology, Sydney in Australia. He had recently spent two months at Complutense University of Madrid, under the programme of Distinguished Visitors from this University.

> **He will be very sadly missed. His passing poses an irreparable loss to the fuzzy logic research field**

Da Ruan chaired the international conference series FLINS on Applied Computational Intelligence from 1994. A Memorial Da Ruan Best Paper Award at each FLINS edition is expected to be acknowledged from now on.

> **He was very energetic, full of brilliant ideas, and always enthusiastic about his research, teaching and collaboration**

A more complete biography, written by his friend and Ph.D. advisor Etienne Kerre, can be found in Fuzzy Sets and Systems 185:1-4 (2011). A Springer book entitled A tribute to Prof. Dr. Da Ruan is expected to appear in 2012.

"Since I have got the sad news about Da my mood has been really down because I have not only lost a brilliant researcher that supported me whenever I needed but also I have lost a close friend that shared me many nice and important moments in my life. I have realized that during more than seven years we have been collaborating Da has been a key factor to my research career. So from these lines I just want to say that I am going to miss him a lot and I will be thankful to him forever. Thanks a lot for being our friend Da."

1.9 Obituary in the Newsletter of Faculty of Engineering and Information Technology, University of Technology, Sydney

Jie Lu

University of Technology Sydney, Broadway, NSW2007, Australia

Adjunct Professor Da Ruan passed away on 31 July, 2011 from heart failure at the early age of 50 years. His death was a shock to his research colleagues and he will be very sadly missed. Da Ruan was a leading member of the global fuzzy logic community. He has been a UTS Adjunct Professor with the DeSI Lab in QCIS since 2006, taking an active part in the Lab's research as PI for two ARC Discovery grants, and as external supervisor for two DeSI PhD students. He also collaborated on 15 joint publications with the DeSI Lab. He is fondly remembered for a seminar in 2006 entitled, "How to Write High Quality Journal Papers".

Adjunct Prof Da Ruan's untimely passing has created much sadness in his research community and poses an irreparable loss to the fuzzy logic research field. He was very energetic, full of brilliant ideas, and always enthusiastic about his research, teaching and collaboration. His close ties with the DeSI Lab kept him in constant contact right up to the time of his early demise.

To honour Da Ruan's memory, the DeSI Lab has created a web page which contains many world-class researchers' messages in tribute to Da's scientific work and his excellent character. He was well liked and highly regarded, as evidenced by the messages of tribute– see: http://decide.it.uts.edu.au/DaRuan/Da_Ruan.html. A short biography of Da Ruan follows below.

Professor Da Ruan's Biography

Professor Da Ruan was a principal research scientist at the Belgian Nuclear Research Centre (SCK•CEN). He received a PhD in Mathematics from Ghent University, Belgium in 1990. His major research interests were in the areas of mathematical modelling, computational intelligence methods, uncertainty analysis, decision support systems and AI applications to information management, cost/benefit analysis, nuclear safety and security related fields. He published 90 peer-reviewed journal articles, two text books, and 20 research books on artificial intelligence, decision support systems, and intelligent information systems and applications.

Dr Ruan chaired the international conference series FLINS on Applied Computational Intelligence from 1994. He received an honorary doctoral degree from the Nuclear Power Institute of China for his research achievements, including intelligent control for nuclear reactors in 1995. He served as a member of the editorial board of *Fuzzy Sets and Systems* (Elsevier), as regional editor for the European *International Journal of Intelligent Automation and Soft Computing* (TSI Press, Albuquerque, NM), as co-editor-in-chief of the *International Journal of Nuclear Knowledge Management* (Interscience Publishers, Geneva), as editor-in-chief of the *International Journal of Computational Intelligence Systems* (Atlantis Press, Paris), and as editor of the book series on *Intelligent Information Systems* and for the proceedings series of *Computer Engineering and Information Science* (World Scientific, Singapore). He was a guest Professor at the Dept of Applied Math and Computer Science in Ghent University and at the Dept of Applied Economics in Hasselt University, respectively, in Belgium, and an Adjunct Professor in the Faculty of Information Technology at University of Technology, Sydney in Australia.

1.10 Obituary in SCK Website: In Memoriam Da Ruan: http://www.sckcen.be/fr/Nouvelles/In-Memoriam-Da-Ruan

With great sadness and unbelief we inform you that our colleague Da Ruan unexpectedly deceased on July 31. The Board of Directors, the management and staff of SCK•CEN express their condolences and deepest sympathy to his family.

Da Ruan began his career at SCK•CEN in 1991, after obtaining his PhD in Mathematics from Ghent University. As a postdoc Da completed the TRANSFUSION project successfully by developing a fast algorithm for the analysis of well-logging signals for the oil industry. Subsequently he took the initiative in the FLINS (Fuzzy Logic In Nuclear Science) project and the FLINS conferences which grew to become the leading conference in this specialized field.

Da Ruan searched tirelessly for applications of Fuzzy Logic and related theories and found them, for example, in reactor control, cost estimates including uncertainties of large projects, decision support systems and the analysis of large data sets for safety culture and the non-proliferation of nuclear weapons. He published more than 90 peer-reviewed journal articles, two text books and 20 research books. His international recognition is also shown by numerous invitations to act as a keynote speaker at international conferences, as well as by the award of an honorary doctorate from the Nuclear Power Institute of China.

Da Ruan served as member of the editorial board of Fuzzy Sets and Systems, as regional editor for Europe of the International Journal of Intelligent Automation and Soft Computing, co-editor-in-chief of the International Journal of Nuclear Knowledge Management, editor-in-chief of the International Journal of Computational Intelligence Systems, editor of the book series Intelligent Information Systems and editor of the proceedings series Computer Engineering and Information Science.

Da Ruan was guest professor at the Department of Applied Mathematics and Computer Science at Ghent University and at the Department of Applied Economics at Hasselt University. At the Faculty of Information Technology at the University of Technology in Sydney, Australia, he was assistant professor.

His colleagues will remember Da as a motivated scientist with an indefatigable but also very congenial personality. With the passing away of Da, SCK•CEN has lost one of its most outstanding and productive scientists. He is survived by his wife Greet and two sons Alexander and Nicolas.

Part II
Da's Scientific Life

2.1 Education

Da was born in Shanghai, China on September 10, 1960. From 1973 to 1979, he studied in Chang Zhi high school (长治中学) in Shanghai. From 1979 to 1983 he studied in Computational Mathematics from the prestigious Fudan University (复旦大学) in Shanghai. After he obtained a Bachelor degree, he worked in Fudan University as an Associate Lecturer and a Master student. In 1987 Da came to Belgium to work towards a PhD under the supervision of Prof. E. Kerre. In 1990 he obtained the PhD degree in Mathematics from Ghent University as well as the Certificate in Management from Leuven University, Belgium.

Da, standing the first from the left, and his French teachers and classmates in Fudan University in 1984

Da Ruan's PhD thesis (1990): A Critical Study of Widely Used Fuzzy Implication Operators and Their Influence on the Inference Rules in Fuzzy Expert Systems. 155p. Ghent University, Belgium

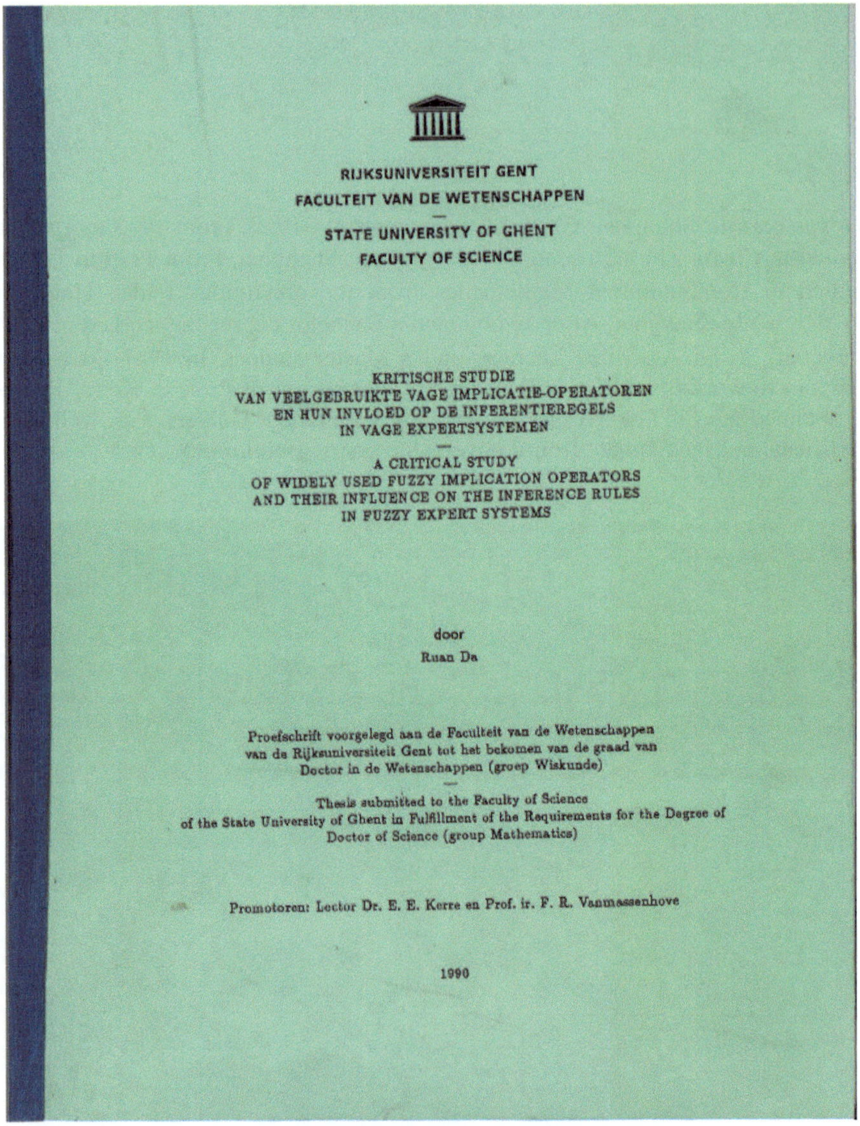

RIJKSUNIVERSITEIT GENT
FACULTEIT VAN DE WETENSCHAPPEN
—
STATE UNIVERSITY OF GHENT
FACULTY OF SCIENCE

KRITISCHE STUDIE
VAN VEELGEBRUIKTE VAGE IMPLICATIE-OPERATOREN
EN HUN INVLOED OP DE INFERENTIEREGELS
IN VAGE EXPERTSYSTEMEN
—
A CRITICAL STUDY
OF WIDELY USED FUZZY IMPLICATION OPERATORS
AND THEIR INFLUENCE ON THE INFERENCE RULES
IN FUZZY EXPERT SYSTEMS

door
Ruan Da

Proefschrift voorgelegd aan de Faculteit van de Wetenschappen
van de Rijksuniversiteit Gent tot het bekomen van de graad van
Doctor in de Wetenschappen (groep Wiskunde)
—
Thesis submitted to the Faculty of Science
of the State University of Ghent in Fulfillment of the Requirements for the Degree of
Doctor of Science (group Mathematics)

Promotoren: Lector Dr. E. E. Kerre en Prof. ir. F. R. Vanmassenhove

1990

Cover page of Da's PhD thesis in 1990

2.2 Scientific Career

Since 1991 till his death he was a permanent staff member of the Belgian Nuclear Research Centre (SCK•CEN) in Mol, Belgium.

2011	April–June: Visiting distinguished professor, Faculty of Mathematics, Complutense University, Madrid (hosted by Profs. B. Vitoriano and J. Montero)
2008	October: Senior visiting scholar, Faculty of Mathematics, Fudan University, Shanghai, China (hosted by Prof. H. Shang and Dr. R. Li)
2008	May: Senior visiting scholar, Faculty of Mathematics, Fudan University, Shanghai, China (hosted by Prof. H. Shang and Dr. R. Li)
2007	July: Visiting scholar, School of Computing and Mathematics, University of Ulster at Jordanstown, Newtownabbey, Co. Antrim, Northern Ireland, UK (hosted by Dr. J. Liu)
2007	March: Visiting scholar, School of Computing and Mathematics, University of Ulster at Jordanstown, Newtownabbey, Co. Antrim, Northern Ireland, UK (hosted by Dr. J. Liu)
2007	January -31 July 2011, Adjunct Professor, University of Technology Sydney, Australia (hosted by Prof. J. Lu)
2006	November: Visiting professor, Faculty of Information Technology, University of Technology Sydney, Australia (hosted by Prof. J. Lu)
2005	December: Visiting professor, Faculty of Information Technology, University of Technology Sydney, Australia (hosted by Prof. J. Lu)
2001	April–September 2002: Visiting scientist, the OECD Halden Reactor Project, Institute for Energy Technology, Halden, Norway
1996–31	July 2011: Full-time senior researcher, project leader, FLINS Chair, SCK•CEN, Mol, Belgium
1994–1995	Post-doc researcher, project leader, SCK•CEN, Mol, Belgium
1991–1993	Principal project researcher, Postdoc, SCK•CEN, Mol, Belgium
1987–1990	PhD candidate in Infinitesimal Analysis Department, Ghent University, Ghent, Belgium (promoter: Prof Dr E.E. Kerre)
1983–1986	Assistant, Department of Mathematics, Fudan University, Shanghai, China

2.3 Research Achievements

2.3.1 Research Experiences

2011: Nuclear safety culture analysis

2010: Security and safety evaluation with intelligent DSS

2008: DSS and AI for energy policy and safeguards research and applications

2006: DSS and AI for safety and security research and applications

2004–2006: Cost/benefits analysis under various uncertainties with AI techniques for large engineering projects (principal investigator)

2003: Machinery intelligence for complex systems (robots, reactors, new control room design for power plants)

2001–2006: Adaptive fuzzy control for a Triga Mark III reactor (cooperation with Mexico) (supported by Mexico)

2001–2002: Development of computational intelligent systems for feedwater flow measurements in nuclear power plants (project leader and principal investigator)

2000: Intelligent systems for industrial engineering (cooperation with Turkey)

2000–2001: Development of a fuzzy-logic-based intelligent tool for uncertainty analysis for nuclear safeguards applications (principal investigator) for IAEA (International Atomic Energy Agency)

1998–1999: Fuzzy control on-line experiments at the BR1 reactor (project leader and principal investigator)

1998: Decision making under uncertainty by intelligent hybrid systems (principal investigator) (cooperation with China)

1996–1998: Fuzzy logic demo model for OMRON Belgium (project leader and principal investigator)

1994–1998: Development of fuzzy controller for the Belgian Reactor 1 (BR1) (project leader and principal investigator)

1992–1993: Fuzzy decision applications in the field of radiation protection

1991–1993: Development of the time-dependent 3-D finite-element diffusion code TRANSFUSION for nuclear oil well logging applications (principal investigator)

1987–1990: PhD on approximate reasoning and expert systems in Ghent University, Gent, Belgium

2.3.2 Grants as a Key Research Scientist

2009–2010: BELSPO Project (12 months), Project no.: c/brs/08
A fuzzy multiple attribute decision-making approach for nuclear safeguards information management (60,000 EUR)

2008–2010: ARC Discovery Project, project no. DP0880739
A Comprehensive Platform for Dynamic Decision Support in Warning Systems through Better Management of Uncertain Information (UTS, Australia with 191,000 AUS$)

2008–2010: BELSPO Project, project no.: SD/EN/07A
Sustainable Energy Policy Integrated Assessment – A normative contribution to decision support (SCK•CEN, Belgium with 199,477 EUR out of a total of 599,414 EUR from other partners in Belgium)

2005–2007: China-Flanders project by Flemish Community, project no.: BIL 011S1105
Intelligent systems for data mining and information processing: methods and applications (Ghent U, Belgium with 50,050 EUR)

2001–2003: The Mexican National Council for Science and Technology (CONACYT) project, no.: 218476-5-33797-A
Adaptive Fuzzy Control and Its Applications in Nuclear Systems (ININ, Mexico with 412,720 MEX $)

2.3.3 Awards and Guest Professorships

2009–2012: Adjunct professor, Shanghai Electricity and Power University, Shanghai, China

2008–2011: Adjunct professor, Shanghai Maritime University, Shanghai, China

2008–2011: Adjunct professor, Shanghai University, Shanghai, China

2007–2010: Guest professor, Faculty of Applied Economics, Hasselt University, Hasselt, Belgium

2007–2012: Adjunct professor, Department of Software, Faculty of Information Technology, University of Technology, Sydney, Australia

2005–2013: Guest professor, Department of Applied Mathematics & Computer Science, Ghent University, Gent, Belgium

2004–2006: Guest professor, Department of Computer Science & Engineering, Xihua University, Chengdu, China

2004–2006: Guest professor, College of Information Science and Technology, Donghua University, Shanghai, China

2001–06: Scientific advisor, the National Institute for Nuclear Research of Mexico, for the project "Adaptive fuzzy control and its applications in nuclear systems," supported by the Mexican National Council for Science and Technology (CONACYT), Mexico

2000– 2011: Advisory professor, Department of Applied Mathematics, Southwest Jiaotong University, Chengdu, Sichuan, China

1995–1998: Concurrent professor, School of Resource and Environmental Science, Beijing Normal University, China

1995: Honorary doctor degree, Nuclear Power Institute of China (NPIC), Chengdu, Sichuan, China

1995: Senior science and technology advisor, NPIC, Chengdu, Sichuan, China

1987–1990: Gent University's scholarship, Gent, Belgium

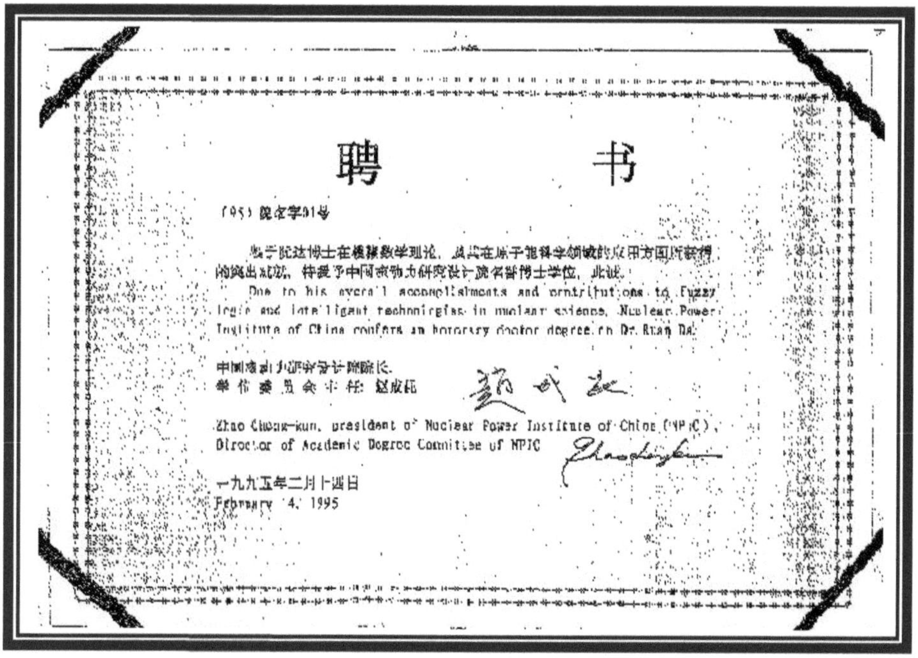
聘　　书

(95) 院学字91号

鉴于阮达博士在模糊数学理论、及其在原子能科学领域的应用方面所获得的突出成就，特授予中国核动力研究设计院名誉博士学位，此证。

Due to his overall accomplishments and contributions to fuzzy logic and intelligent technologies in nuclear science, Nuclear Power Institute of China confers an honorary doctor degree on Dr. Ruan Da.

中国核动力研究设计院院长
学位委员会主任：赵成昆

Zhao Cheng-kun, president of Nuclear Power Institute of China (NPIC),
Director of Academic Degree Committee of NPIC

一九九五年二月十四日
February 14, 1995

Da received an honorary doctor degree from the Nuclear Power Institute of China in 1995

199[illegible] Scientific and technological advisor, [illegible] Government, [illegible], China

198[illegible]–199[illegible] [illegible]

2.4 Publications

2.4.1 Books, Edited Volumes and Proceedings

1. Jin X, Liu Y, Li T, Ruan D (2010) Proceedings of the 2010 International Conference on Intelligent Systems and Knowledge Engineering (ISKE2010). IEEE Press, Beijing, China

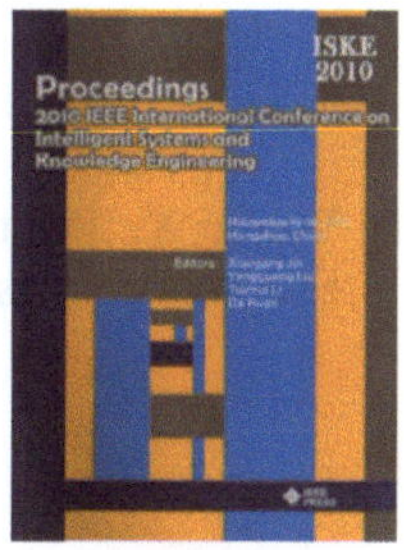

2. Ruan D, Li T, Xu Y, Chen G, Kerre EE (2010) Proceedings of FLINS2010 on Computational Intelligence: Foundations and Applications. World Scientific, Singapore

3. Ruan D (2010) Computational Intelligence in Complex Decision Systems. Atlantis Computational Intelligence System Book Series, Atlantis Press/World Scientific, Paris/Singapore

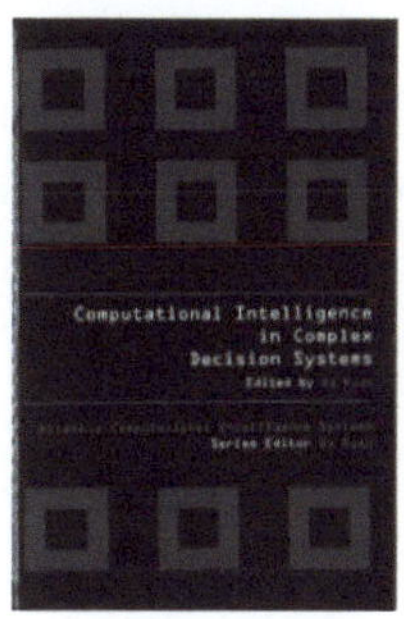

4. Pei Z, Ruan D, Liu J, Xu Y (2009) Linguistic Valued Based Intelligent Information Processing: Theory, Methods, and Applications. Atlantis Press/World Scientific, Paris/Singapore

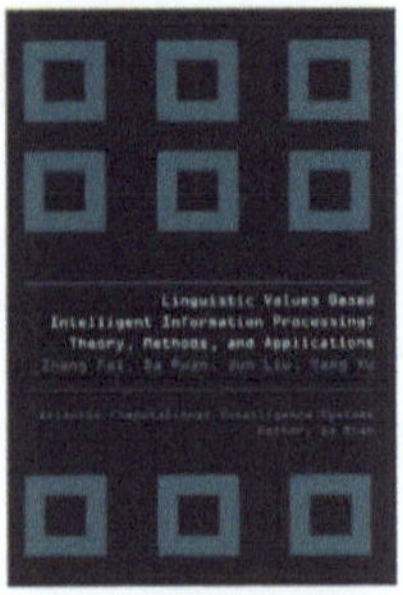

5. Vanhoof K, Ruan D, Li T, Wets G (2009) Proceedings of the 2009 International Conference on Intelligent Systems and Knowledge Engineering (ISKE2009): Intelligent Decision Making Systems. World Scientific, Singapore

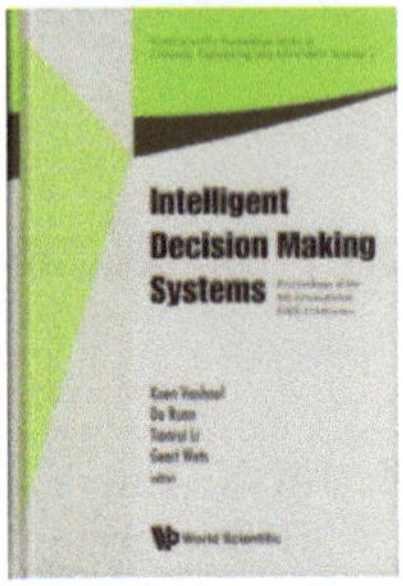

6. Wang X, Ruan D, Kerre EE (2009) Mathematics of Fuzziness – Basic Issues. Springer, Heidelberg

7. Li S, Li T, Ruan D (2008) Proceedings of the 2008 International Conference on Intelligent Systems and Knowledge Engineering (ISKE2008). IEEE Press, Beijing, China

8. Ruan D, Montero J, Lu J, Martínez L, D'hondt P, Kerre EE (2008) Proceedings of FLINS2008 on Computational Intelligence in Decision and Control. World Scientific, Singapore

9. Ruan D, Hardeman F, Van der Meer K (2008) Intelligent Decision and Policy Making Support Systems. Studies in Computational Intelligence. Springer, Heidelberg

10. Wang PP, Ruan D, Kerre EE (2007) Fuzzy Logic – A Spectrum of Theoretical & Practical Issues. Studies in Fuzziness and Soft Computing. Springer, Heidelberg

11. Lu J, Zhang G, Ruan D, Wu F (2007) Multi-Objective Group Decision Making – Methods, Software and Applications with Fuzzy Set Technology. Imperial College Press, London

12. Li T, Xu Y, Ruan D (2007) Proceedings of the 20071. International Conference on Intelligent Systems and Knowledge Engineering (ISKE2007). Atlantis Press, Paris

13. Zeng X, Li Y, Ruan D, Koehl L (2007) Computational Textile. Springer, Heidelberg

14. Lu J, Ruan D, Zhang G (2007) E-Service Intelligence. Studies in Computational Intelligence. Springer, Heidelberg

15. Ruan D, D'hondt P, Fantoni PF, De Cock M, Nachtegael M, Kerre EE (2006) Proceedings of FLINS 2006 on Applied Artificial Intelligence. World Scientific, Singapore

16. Ruan D, Chen G, Kerre EE, Wets G (2005) Intelligent Data Mining. Studies in Computational Intelligence. Springer, Heidelberg

17. Kerre EE, Huang C, Ruan D (2004) Fuzzy Sets Theory and Approximate Reasoning (in Chinese). Wuhan University Press, Wuhan

18. Ruan D, D'hondt P, De Cock M, Nachtegael M, Kerre EE (2004) Proceedings of FLINS 2004 on Applied Computational Intelligence. World Scientific, Singapore

19. Ruan D, Zeng X (2004) Intelligent Sensory Evaluation. Springer, Heidelberg

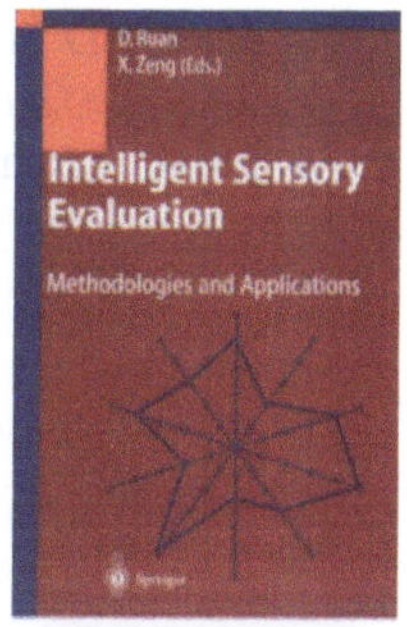

20. Xu Y, Ruan D, Qin K, Liu J (2003) Lattice-Valued Logic — An Alternative Approach to Treat Fuzziness and Incomparability. Springer, Heidelberg

21. Zhou C, Maravall D, Ruan D (2003) Autonomous Robotic Systems — Soft Computing and Hard Computing Methodologies and Applications. Physica-Verlag, Heidelberg

22. Ruan D, Fantoni PF (2002) Power Plant Surveillance and Diagnostics — Applied Research with Artificial Intelligence. Springer, Heidelberg

23. Ruan D, D'hondt P, Kerre EE (2002) Proceedings of FLINS 2002 on Computational Intelligent Systems for Applied Research. World Scientific, Singapore

24. Ruan D, Kacprzyk J, Fedrizzi M (2001) Soft Computing in Risk Evaluation and Management — Applications in Technology, Environment and Finance. Physica-Verlag, Heidelberg

25. Ruan D, Aït Abderrahim H, D'hondt P, Kerre EE (2000) Proceedings of FLINS 2000 on Intelligent Technologies and Soft Computing in Nuclear Science and Engineering. World Scientific, Singapore

26. Ruan D, Kerre EE (2000) Fuzzy If-Then Rules in Computational Intelligence: Theory and Applications. Kluwer Academic Publishers, Boston

27. Ruan D, Huang C (2000) Fuzzy Sets and Fuzzy Information-Granulation Theory – Key Selected Papers by Lotfi Zadeh. Beijing Normal University Press, Beijing (both in English and in Chinese)

28. Ruan D, Huang C (2000) Fuzzy Sets and Operations Research for Decision Support – Key Selected Papers by Hans-Jürgen Zimmermann. Beijing Normal University Press, Beijing

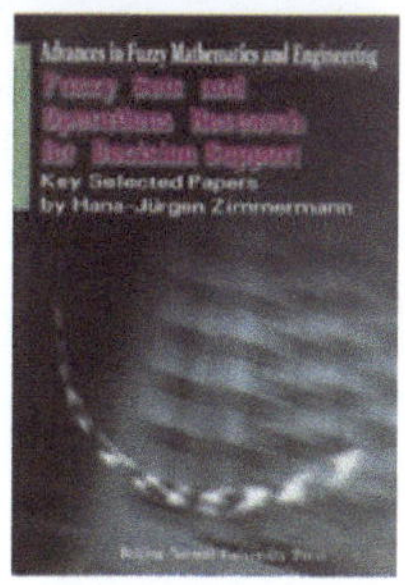

29. Ruan D (1999) Fuzzy Systems and Soft Computing in Nuclear Engineering. Physica-Verlag, Heidelberg

30. Ruan D, Aït Abderrahim H, D'hondt P, Kerre EE (1998) Proceedings of FLINS1998 onFuzzy Logic and Intelligent Technologies in Nuclear Science and Industry. World Scientific, Singapore

31. Ruan D (1997) Intelligent Hybrid Systems: Fuzzy Logic, Neural Networks, and Genetic Algorithms. Kluwer Academic Publishers, Boston, USA

32. Ruan D (1996) Fuzzy Logic Foundations and Industrial Applications. Kluwer Academic Publishers, Boston, USA

33 Ruan D, D'hondt P, Govaerts P, Kerre EE (1996) Proceedings of FLINS1996 on Intelligent Systems and Soft Computing for Nuclear Science and Industry. World Scientific, Singapore

34 Ruan D (1995) Fuzzy Set Theory and Advanced Mathematical Applications. Kluwer Academic, Boston, USA

35 De Cooman G, Ruan D, Kerre EE (1995) Foundations and Applications of Possibility Theory, Proceedings of FAPT'95. World Scientific, Singapore

36 Ruan D, D'hondt P, Govaerts P, Kerre EE (1994) Proceedings of the 1994 International Conference on Fuzzy Logic and Intelligent Technologies in Nuclear Science (FLINS1994), World Scientific, Singapore

2.4.2 Journal Papers

1. Liu, J., Chen, S.W., Xu, Y., Ruan, D., Martínez, L.: Axiomatizable lattice-ordered qualitative linguistic truth-valued logic. submitted to International Journal of General Systems (2011) (under review)
2. Kabak, Ö., Ruan, D.: A cumulative belief degree-based approach for missing values in nuclear safeguards evaluation. IEEE Transactions on Knowledge and Data Engineering (2011) (accepted, in press)
3. He, X.X., Xu, Y., Liu, J., Ruan, D.: α -Lock resolution method for lattice-valued logic based on lattice implication algebra. Engineering Applications of Artificial Intelligence (accepted, in press) (2011)
4. He, X.X., Liu, J., Xu, Y., Martínez, L., Ruan, D.: On α-satisfiability and its α-Lock resolution in a finite lattice-valued logic. Logic Journal of the IGPL (2011), doi:10.1093/jigpal/JZR007
5. Kabak, O., Ruan, D.: A comparison study of fuzzy MADM methods in nuclear safeguards evaluation. Journal of Global Optimization 51, 209–226 (2011), doi:10.1007/s10898-010-9601-1
6. Liu, D., Li, T., Ruan, D., Zhang, J.: Incremental learning optimization on knowledge discovery in dynamic business intelligent systems. Journal of Global Optimization 51, 325–344 (2011), doi:10.1007/s10898-010-9607-8
7. Liu, D., Li, T., Ruan, D.: Probabilistic model criteria with decision-theoretic rough sets. Information Sciences 181(17), 3709–3722 (2011)
8. Liu, J., Martínez, L., Ruan, D., Rodriguez, R., Calzada, A.: Optimization algorithm for learning consistent belief rule-base from examples. Journal of Global Optimization 51(2), 255–270 (2011), doi:10.1007/s10898-010-9605-x
9. Rodriguez, R.M., Ruan, D., Liu, J., Martínez, L.: A hybrid model to deal with missing values in nuclear safeguards' evaluation. International Journal of Nuclear Knowledge Management 5(2), 194–218 (2011)
10. Shen, Y., Ruan, D., Hermans, E., Brijs, T., Wets, G., Vanhoof, K.: Modeling qualitative data in data envelopment analysis for composite indicators. International Journal of Systems Assurance Engineering and Management 2(1), 21–30 (2011), doi:10.1007/s13198-011-0051-z
11. Xu, Y., Liu, J., Ruan, D., Li, X.: Determination of a-resolution in lattice-valued first-order logic LF(X). Information Sciences 181, 1836–1862 (2011)
12. Zou, L., Ruan, D., Pei, Z., Xu, Y.: A linguistic-valued lattice implication algebra approach for risk analysis. Journal of Multiple Valued Logic and Soft Computing 17(4), 293–303 (2011)
13. Martínez, L., Ruan, D., Herrera, F.: Computing with words in decision support systems: an overview on models and applications. International Journal of Computational Intelligence Systems 3(4), 382–395 (2010), doi:10.2991/ijcis.2010.3.4.1
14. Zhu, Y., Ruan, D., Zeng, X., Koehl, L.N., Chaigneau, C.: Characterization of fashion themes using fuzzy techniques for designing new human centered products. International Journal of Computational Intelligence Systems 3(4), 452–460 (2010), doi:10.2991/ijcis.2010.3.4.5
15. Ruan, D.: Lessons learned from soft computing applications at SCK-CEN. International Journal of Computational Intelligence Systems 3(2), 249–255, doi:10.2991/ijcis, 3.2.12 (2010)

16. Chen, H., Li, T., Qiao, S., Ruan, D.: A rough set based dynamic maintenance approach for approximations in coarsening and refining attribute values. International Journal of Intelligent Systems 25(10), 1005–1026 (2010)
17. Rodriguez, R.M., Martínez, L., Ruan, D., Liu, J.: Using collaborative filtering for dealing with missing values in nuclear safeguards evaluation. International Journal of Uncertainty, Fuzziness and Knowledge-Based Systems 18(4), 431–449 (2010), doi:10.1142/S0218488510006635
18. Ruan, D., Lu, J., Laes, E., Zhang, G., Ma, J., Meskens, G.: Multi-criteria group decision support with linguistic variables in long-term scenarios for Belgian energy policy. Journal of Universal Computer Science 16(1), 103–120 (2010)
19. Xu, Y., Lu, J., Martínez, L., Ruan, D.: Some views on information fusion and logic based approaches in decision making under uncertainty. Journal of Universal Computer Science 16(1), 3–21 (2010)
20. Shen, Y., Hermans, E., Ruan, D., Wets, G., Brijs, T., Vanhoof, K.: Evaluating trauma management performance in Europe. Transportation Research Record: Journal of the Transportation Research Board 2(2148), 69–75 (2010)
21. Büyüközkan, G., Ruan, D.: Choquet integral based aggregation approach to software development risk assessment. Information Sciences 18(3), 441–451 (2010), doi:10.1016/j.ins.2009.09.009
22. Shen, Y., Li, T., Hermans, E., Ruan, D., Wets, G., Vanhoof, K., Brijs, T.: A hybrid system of neural networks and rough sets for road safety performance indicators. Soft Computing 14(12), 1255–1263 (2010), doi:10.1007/s00500-009-0492-3
23. Hermans, E., Ruan, D., Brijs, T., Wets, G., Vanhoof, K.: Road safety risk evaluation by means of ordered weighted averaging operators and expert knowledge. Knowledge Based Systems 23(1), 48–52 (2010)
24. Shi, Y., Nachtegael, M., Ruan, D., Kerre, E.E.: Fuzzy adjunctions and fuzzy morphological operations based on implications. International Journal of Intelligent Systems 24(12), 1289–1296 (2009)
25. Liu, D., Li, T., Ruan, D., Zou, W.: An incremental approach for inducing knowledge from dynamic information systems. Fundamenta Informaticae 94(2), 245–260 (2009)
26. Song, J., Li, T., Ruan, D.: An integration of cloud transform and rough set theory to induction of decision trees. Fundamenta Informaticae 94(2), 261–273 (2009)
27. Li, X., Chen, D., Ruan, D., Tang, B.: Evaluation models of insurers' risk management based on large system theory. Stochastic Environmental Research and Risk Assessment 23(4), 415–423 (2009)
28. Liu, J., Ruan, D., Wang, H., Martinez, L.: Improving nuclear safeguards evaluation through enhanced belief rule-based inference methodology. International Journal of Nuclear Knowledge Management 3(3), 312–339 (2009)
29. Pei, Z., Xu, Y., Ruan, D., Qin, K.: Extracting complex linguistic data summaries from personnel database via simple linguistic aggregations. Information Sciences 179(14), 2325–2332 (2009)
30. Zou, L., Ruan, D., Pei, Z., Xu, Y.: A linguistic truth-valued reasoning approach in decision making with incomparable information. Journal of Intelligent and Fuzzy Systems 19(4/5), 335–343 (2008)
31. Shi, Y., Van Gasse, B., Ruan, D., Kerre, E.E.: On the first place antitonicity in QL-implications. Fuzzy Sets and Systems 159, 2988–3013 (2008)
32. Huang, C., Ruan, D.: Fuzzy risks and an updating algorithm with new observations. Risk Analysis 28(3), 681–694 (2008)

33. Liu, J., Yang, J.B., Ruan, D., Martínez, L., Wang, J.: Self-tuning of fuzzy belief rule bases for engineering system safety analysis. Annals of Operations Research 163(1), 143–168 (2008)
34. Zeng, X., Ruan, D., Koehl, L.: Intelligent sensory evaluation: concepts, implementations, and applications. Mathematics and Computers in Simulation 77(5-6), 443–452 (2008)
35. Büyüközkan, G., Ruan, D.: Evaluation of software development projects using a fuzzy multi-criteria decision approach. Mathematics and Computers in Simulation 77(5-6), 464–474 (2008)
36. Xu, D.L., Yang, J.B., Carlé, B., Hardeman, F., Ruan, D.: Application of an intelligent decision system to nuclear waste repository option analysis. International Journal of Nuclear Governance, Economy and Ecology 2(2), 145–165 (2008)
37. Lu, J., Zhang, G., Ruan, D.: Intelligent multi-criteria fuzzy group decision-making for situation assessments. Soft Computing 12(3), 289–299 (2008)
38. Ruan, D., Li, T.: Uncertainty analysis via aggregation operators for cost estimation in a larger scale nuclear project. International Journal of Nuclear Knowledge Management 2(4), 375–395 (2007)
39. Benítez-Read, J.S., Rivero-Gutiérrez, T., Ruan, D., Ramírez-Chávez, C.L., López-Callejas, R., Pacheco-Sotelo, J.O.: User interface for intelligent control schemes in a TRIGA Mark III reactor. International Journal of Nuclear Knowledge Management 2(3), 268–284 (2007)
40. Li, T., Ruan, D., Wets, G., Song, J., Xu, Y.: A rough sets based characteristic relation approach for dynamic attribute generalization in data mining. Knowledge Based Systems 20(5), 485–494 (2007)
41. Shi, Y., Ruan, D., Kerre, E.E.: On the characterizations of fuzzy implications satisfying I(x,y)=I(x,I(x,y)). Information Sciences 177(14), 2954–2970 (2007)
42. Li, T., Ruan, D.: An extended process model of knowledge discovery in database. Journal of Enterprise Information Management 20(2), 169–177 (2007)
43. Ding, Y., Gao, L., Ruan, D.: Communication mechanisms in ecological network-based grid middleware for service emergence. Information Sciences 177, 722–733 (2007)
44. Pei, Z., Ruan, D., Xu, Y., Liu, J.: Handling linguistic information over the internet based on a multi-agent system. International Journal of Intelligent Systems 22(5), 435–453 (2007)
45. Büyüközkan, G., Ruan, D., Feyzioğlu, O.: Evaluating e-learning web site quality in a fuzzy environment. International Journal of Intelligent Systems 22(5), 567–586 (2007)
46. Ma, J., Li, W., Ruan, D., Xu, Y.: Filter-based resolution principle for lattice-valued propositional logic LP(X). Information Sciences 177, 1046–1062 (2007)
47. Büyüközkan, G., Feyzioğlu, O., Ruan, D.: Fuzzy group decision-making to multiple preference formats in quality function deployment. Computers in Industry 58(5), 392–402 (2007)
48. Ma, J., Ruan, D., Xu, Y., Zhang, G.: A fuzzy-set approach to treat determinacy and consistency of linguistic terms in multi-criteria decision making. International Journal of Approximate Reasoning 44, 165–181 (2007)
49. Martínez, L., Liu, J., Ruan, D., Yang, J.B.: Dealing with heterogeneous information in engineering evaluation processes. Information Sciences 177, 1533–1542 (2007)

50. Lu, J., Shi, C., Zhang, G., Ruan, D.: An extended Branch and Bound algorithm for bilevel multi-follower decision making in a referential-uncooperative situation. International Journal of Information Technology and Decision Making 6(2), 371–388 (2007)
51. Cebeci, U., Ruan, D.: A multi-attribute comparison of Turkish quality consulants by fuzzy AHP. International Journal of Information Technology & Decision Making (IT&DM) 6(1), 191–207 (2007)
52. Liu, X., Ruan, D., Li, T.: A credit evaluation in E-business through game theory. New Mathematics and Natural Computation 3(1), 91–109 (2007)
53. Lu, J., Ruan, D., Wu, F., Zhang, G.: An α-fuzzy goal approximate algorithm for solving fuzzy multiple objective linear programming problems. Soft Computing 11(3), 259–267 (2007)
54. Liang, X., Liu, X., Ruan, D., Zhu, D., Tang, B.: An adaptive T-S fuzzy control model for the pension information systems. Dynamics of Continuous, Discrete & Impulsive Systems, Series B: Applications & Algorithms 13(b), 106–110 (2006)
55. Shi, Y., Zhang, L., Ruan, D., Cai, K., Kerre, E.E.: On the robustness of fuzzy connectives in control systems. Dynamics of Continuous, Discrete & Impulsive Systems, Series B: Applications & Algorithms 13(b), 121–125 (2006)
56. Büyüközkan, G., Ruan, D., Feyzioğlu, O.: Effective software process management in quality function deployment by a fuzzy-logic based group decision-making approach. Discrete & Impulsive Systems, Series B: Applications & Algorithms 13(b), 141–145 (2006)
57. Xu, Y., Liu, J., Ruan, D., Lee, T.T.: On the consistency of rule bases based on lattice-valued first-order logic LF(X). International Journal of Intelligent Systems 21(4), 399–424 (2006)
58. Ertay, T., Ruan, D., Tuzkaya, U.R.: Integrating data envelopment analysis and analytic hierarchy for the facility layout design in manufacturing systems. Information Sciences 176(6), 237–262 (2006)
59. Bilsel, U., Büyüközkan, G., Ruan, D.: A fuzzy preference-ranking model for a quality evaluation of hospital web sites. International Journal of Intelligent Systems 21, 1181–1197 (2006)
60. Ruan, D., Benitez-Read, J.S.: Fuzzy control for nuclear reactor operation – strengths, weaknesses, opportunities and threats. Journal of Intelligent and Fuzzy Systems 16(4), 289–295 (2005)
61. Ertay, T., Büyüközkan, G., Kahraman, C., Ruan, D.: Quality function deployment implementation based on analytic network process with linguistic data: an application in automotive industry. Journal of Intelligent and Fuzzy Systems 16(3), 221–232 (2005)
62. Benitez-Read, J.S., Ruan, D.: International cooperation on control for safe operation of nuclear research reactors. Progress in Nuclear Energy 46(3-4), 309–320 (2005)
63. Benitez-Read, J.S., Ruan, D., Najera-Hernanbez, M., Pérez-Clavel, B., Pacjeco-Sotelo, J.O., Lopez-Callejas, R.: Comparison between a continuous and a discrete method for the aggregation and defuzzification stages of a TRIGA reactor power fuzzy controller. Progress in Nuclear Energy 46(3-4), 321–327 (2005)
64. Ertay, T., Ruan, D.: Data envelopment analysis based decision model for optimal operator allocation in CMS. European Journal of Operational Research 164(3), 800–810 (2005)

65. Liu, X., Ruan, D., Xu, Y.: A study of enterprise human resource competence appraisement. International Journal of Enterprise Information Management 18(3), 289–315 (2005)
66. Gülbay, M., Kahraman, C., Ruan, D.: α-Cut fuzzy control charts for linguistic data. International Journal of Intelligent Systems 19(12), 1173–1195 (2004)
67. Kahraman, C., Bozdag, C.E., Ruan, D., Ozok, A.F.: Fuzzy sets approaches to statistical parametric and nonparametric tests. International Journal of Intelligent Systems 19, 1069–1087 (2004)
68. Ruan, D.: The role of computational intelligence in sensory evaluation. Journal of Donghua University, Shanghai 3(3), 21–25 (2004)
69. Xu, Y., Zeng, X., Ruan, D.: On normalization of comprehensive evaluation system with complex uncertain information. Journal of Donghua University, Shanghai 3(3), 30–35 (2004)
70. Büyüközkan, G., Ertay, T., Kahrama, C., Ruan, D.: Determining the importance weights for the design requirements in house of quality using fuzzy analytic network approach. International Journal of Intelligent Systems 19(5), 443–461 (2004)
71. Roverso, D., Ruan, D.: Enhancing cross-correlation analysis with artificial neural networks for nuclear power plant feedwater flow measurement. Real-Time Systems 27(1), 85–96 (2004)
72. Büyüközkan, G., Kahraman, C., Ruan, D.: A fuzzy multi-criteria decision approach for software development strategy selection. International Journal of General Systems 33(2-3), 259–280 (2004)
73. Kahraman, C., Beskese, A., Ruan, D.: Measuring flexibility of computer integrated manufacturing systems using fuzzy cash flow analysis. Information Sciences 168, 77–94 (2004)
74. Kahraman, C., Cebeci, U., Ruan, D.: Multi-attribute comparison of catering service companies using fuzzy AHP. International Journal of Production Economics 87, 171–184 (2004)
75. Ruan, D.: Initial experiments on fuzzy control for nuclear reactor operations at the BR1. Nuclear Technology 143(2), 227–240 (2003)
76. Ruan, D., Roverso, D., Fantoni, P.F., Sanabrias, J.I., Carrasco, J.A., Fernandez, L.: Integrating cross- correlation techniques and neural networks for feedwater flow measurement. Progress of Nuclear Energy 43(1-4), 267–274 (2003)
77. Ruan, D., Liu, J., Carchon, R.: Linguistic assessment approach for managing nuclear safeguards indicator information. International Journal of Logistics Information Management 16(6), 401–419 (2003)
78. Bozdag, C.E., Kahraman, C., Ruan, D.: Fuzzy group decision making for selection among computer integrated manufacturing systems. Computers in Industry 51, 13–29 (2003)
79. Kahraman, C., Ruan, D., Doğan, I.: Fuzzy group decision-making for facility location selection. Information Sciences 157, 135–153 (2003)
80. Liu, J., Ruan, D., Xu, Y., Song, Z.: A resolution-like strategy based on a Lattice-valued logic. IEEE Transactions on Fuzzy Systems 11(4), 560–567 (2003)
81. Xu, Y., Liu, J., Ruan, D.: Rule acquisition and adjustment based on set-valued mapping. Information Sciences 157, 167–198 (2003)
82. Kahraman, C., Ruan, D., Bozdag, C.E.: Optimization of multilevel investments using dynamic programming based on fuzzy cash flows. Fuzzy Optimization and Decision Making Journal 2(2), 101–122 (2003)

83. Liu, J., Ruan, D., Carchon, R.: Synthesis and evaluation analysis of the indicator information in nuclear safeguards applications by computing with words. International Journal of Applied Mathematics and Computer Science 12(3), 449–462 (2002)
84. Ruan, D., Fantoni, P.F.: Introduction: intelligent systems for process monitoring and diagnosis. International Journal of Intelligent Systems 17(8), 717–721 (2002)
85. Xu, Y., Liu, J., Ruan, D., Li, W.: Fuzzy reasoning based on generalized fuzzy if-then rules. International Journal of Intelligent Systems 17(10), 977–1006 (2002)
86. Ruan, D.: Aggregation operators for engineering applications. International Journal of Computer Research 11(2), 145–157 (2002)
87. Kahraman, C., Ruan, D., Tolga, E.: Capital budgeting techniques using discounted fuzzy versus probabilistic cash flows. Information Sciences 142(1-4), 57–76 (2002)
88. Zhou, C., Ruan, D.: Fuzzy control rules extraction from perception-based information using computing with words. Information Sciences 142(1-4), 275–290 (2002)
89. Ruan, D.: Implementation of adaptive fuzzy control for a real time control demo-model. Real-Time Systems 21, 219–239 (2001)
90. Xu, Y., Kerre, E.E., Ruan, D., Song, Z.: Fuzzy reasoning based on the extension principle. International Journal of Intelligent Systems 16, 469–495 (2001)
91. Xu, Y., Ruan, D., Kerre, E.E., Liu, J.: α-Resolution principle based on lattice-valued propositional logic LP(X). Information Sciences 130, 195–223 (2000)
92. Ruan, D.: Safety regulations and fuzzy-logic control to nuclear reactors. Mathware & Soft Computing VII (3), 351–360 (2000)
93. Li, X., Ruan, D.: Comparative study of fuzzy control, PID control, and advanced fuzzy control for simulating a nuclear reactor operation. International Journal of General Systems 29(2), 263–279 (2000)
94. Li, X., Ruan, D.: Novel neural algorithms based on fuzzy δ rules for solving fuzzy relation equations: Part 3. Fuzzy Sets and Systems 109(3), 355–362 (2000)
95. Zhou, C., Ruan, D.: Integration of linguistic and numerical information for biped control. Robotics and Autonomous Systems 28(1), 53–70 (1999)
96. Li, X., Ruan, D.: Novel neural algorithms based on fuzzy δ rules for solving fuzzy relation equations: Part 2. Fuzzy Sets and Systems 103(3), 473–486 (1999)
97. Ruan, D., Van der Wal, A.J.: Controlling the power output of a nuclear reactor with fuzzy logic. Information Sciences 110, 151–177 (1998)
98. Ruan, D., Li, X.: Fuzzy-logic control applications to the Belgian Nuclear Reactor 1 (BR1). Computers and Artificial Intelligence 17(2-3), 127–150 (1998)
99. Li, X., Ruan, D.: Fuzzy δ rule and its simulations in fuzzy relation equations. The Journal of Fuzzy Mathematics 6(3), 781–792 (1998)
100. Ruan, D.: Intelligent systems in nuclear applications. International Journal of Intelligent Systems 13(2/3), 115–125 (1998)
101. Resconi, G., Van der Wal, A.J., Ruan, D.: Speed-up of the Monte Carlo method by using a physical model of the Dempster-Shafer theory. International Journal of Intelligent Systems 13(2/3), 221–242 (1998)
102. Li, X., Ruan, D., Van der Wal, A.J.: Discussion on soft computing at FLINS'96. International Journal of Intelligent Systems 13(2/3), 287–300 (1998)
103. Ruan, D., Li, X.: The potential of fuzzy logic in nuclear reactor control. Nouvelles de la Science et des Technologies, Bruxelles, N.S.T. 15(2/3/4), 99–102 (1997)
104. Zhou, C., Ruan, D., Zhu, S.: Hybrid intelligent fuzzy control for nonlinear systems. Journal of Communication & Cognition 14(3), 163–189 (1997)

105. Ruan, D.: Fuzzy logic in nuclear engineering. Belgian Journal of Operations Research, Statistics and Computer Science 37(1-2), 29–45 (1997)
106. Li, X., Ruan, D.: Novel neural algorithms based on fuzzy δ rules for solving fuzzy relation equations: Part 1. Fuzzy Sets and Systems 90(1), 11–23 (1997)
107. Ruan, D.: Fuzzy logic in nuclear safety issue. Journal of Fuzzy Logic and Intelligent Systems 3(1), 34–44 (1995)
108. Ruan, D.: Fuzzy logic in the nuclear research world. Fuzzy Sets and Systems 74(1), 5–13 (1995)
109. Ruan, D., Van Den Durpel, L., Yin, X., D'hondt, P.: Fuzzy logic control for a nuclear reactor (invited) in the Special issue of Journal A on Fuzzy Control 36(3), 62–66 (1995)
110. Ruan, D., De Raedt, C., D'hondt, P., Mund, E.: TRANSFUSION: A time-dependent finite-element code for the solution of the diffusion equation in nuclear oil well logging applications. International Journal of Mathematical Modelling and Scientific Computing 6 (1996)
111. De Cooman, G., Ruan, D., Ryjov, A.P.: FLINS-related activities in Russia. Fuzzy Sets and Systems 74(1), 163–173 (1995)
112. Wang, X., Kerre, E.E., Ruan, D.: Consistency of judgment matrix and fuzzy weights in fuzzy analytic hierarchy process. International Journal of Uncertainty, Fuzziness and Knowledge-Based Systems 3(1), 35–46 (1995)
113. Wang, X., Cappelle, B., Kerre, E.E., Ruan, D.: Transitivity of fuzzy ordering based on pairwise comparisons. The Journal of Fuzzy Mathematics 3(2), 455–463 (1995)
114. Ruan, D., Kerre, E.E.: Fuzzy implication operators and generalized method of cases. Fuzzy Sets and Systems 54(1), 23–38 (1993)
115. Ruan, D., Kerre, E.E.: On the extension of the compositional rule of inference. International Journal of Intelligence Systems 8, 807–817 (1993)
116. Sohier, A., Van Camp, M., Ruan, D., Govaerts, P.: Methods for radiological assessment in the near-field during the early phase of an accidental release of radioactive material using an incomplete data base. Radiation Protection Dosimetry 50(2-4), 321–325 (1993)
117. Van de Walle, B., Ruan, D., Kerre, E.E.: Applications of fuzzy reasoning in nuclear decision aiding systems. International Journal of Fuzzy Systems & Artificial Intelligence 2(2), 35–46 (1993)
118. Cappelle, B., Kerre, E.E., Ruan, D., Vanmassenhove, F.: Characterization of binary operations on the unit interval satisfying the generalized Modus Ponens inference rule. Mathematica Pannonica 2(1), 105–121 (1991)
119. Tang, H., Debaveye, J., Ruan, D., Van Ranst, E.: Land suitability classification based on fuzzy set theory. Pedologie XLI-3, 277–290 (1991)
120. Ruan, D., Kerre, E.E., De Cooman, G., Cappelle, B., Vanmassenhove, F.: Influence of the fuzzy implication operator on the method-of-cases inference rule. International Journal of Approximate Reasoning 4, 307–318 (1990)
121. De Cooman, G., Kerre, E.E., Cappelle, B., Ruan, D., Vanmassenhove, F.: On the extension of classical propositional logic by means of a triangular norm. International Journal of Intelligent Systems 5(3), 307–322 (1990)

2.4.3 Book Chapters

1. Laes, E., Meskens, G., Ruan, D., Lu, J., Zhang, G., Wu, F., D'haesleer, W., Weiler, R.: Fuzzy-set decision support for a Belgian long-term sustainable energy strategy. In: Ruan, D., Hardeman, F., Van der Meer, K. (eds.) Intelligent Decision and Policy Making Support Systems. SCI, vol. 117, pp. 271–296. Springer, Heidelberg (2008)
2. Wang, P.P., Ruan, D., Kerre, E.E.: Why fuzzy logic? – a spectrum of theoretical and pragmatics issues. In: Wang, P.P., Ruan, D., Kerre, E.E. (eds.) Fuzzy Logic. STUDFUZZ, vol. 215, pp. 1–13. Springer, Heidelberg (2007)
3. Lu, J., Ruan, D., Zhang, G.: E-service intelligence: an introduction. In: Lu, J., Ruan, D., Zhang, G. (eds.) Service Intelligence. SCI, vol. 37, pp. 1–33. Springer, Heidelberg (2007)
4. Li, T., Xu, Y., Ruan, D., Pan, W.: Sequential pattern mining. In: Ruan, D., Chen, G., Kerre, E.E., Wets, G. (eds.) Intelligent Data Mining–Techniques and Applications. SCI, vol. 5, pp. 103–122. Springer, Heidelberg (2005)
5. Tocatlidou, A., Ruan, D., Kaloudis, S.T., Lorentzos, N.A.: Uncertain knowledge association through information gain. In: Ruan, D., Chen, G., Kerre, E.E., Wets, G. (eds.) Intelligent Data Mining – Techniques and Applications. SCI, vol. 5, pp. 123–135. Springer, Heidelberg (2005)
6. Tuzkaya, U.R., Ertay, T., Ruan, D.: Simulated annealing approach for the multi-objective facility layout problem. In: Ruan, D., Chen, G., Kerre, E.E., Wets, G. (eds.) Intelligent Data Mining – Techniques and Applications. SCI, vol. 5, pp. 417–418. Springer, Heidelberg (2005)
7. Liu, J., Ruan, D., Yang, J., Martínez-Lopez, L.: Self-tuning fuzzy rule bases with belief structure. In: Ruan, D., Chen, G., Kerre, E.E., Wets, G. (eds.) Intelligent Data Mining – Techniques and Applications. SCI, vol. 5, pp. 417–418. Springer, Heidelberg (2005)
8. Ruan, D., Zeng, X.: Intelligent sensory evaluation: an introduction. In: Ruan, D., Zeng, X. (eds.) Intelligent Sensory Evaluation – Methodologies and Applications, pp. 1–9. Springer, Heidelberg (2004)
9. Resconi, G., Ruan, D.: Multidimensional spaces of objects and features towards intelligent sensory fusion. In: Ruan, D., Zeng, X. (eds.) Intelligent Sensory Evaluation – Methodologies and Applications, pp. 13–29. Springer, Heidelberg (2004)
10. Kahraman, C., Büyüközkan, G., Ruan, D.: Measuring software development value using fuzzy logic. In: Ruan, D., Zeng, X. (eds.) Intelligent Sensory Evaluation – Methodologies and Applications, pp. 285–308. Springer, Heidelberg (2004)
11. Laes, E., Meskens, G., Ruan, D.: Planning for sustainability in the Belgian electricity sector: a multi-criteria mapping exercise. In: Ruan, D., Zeng, X. (eds.) Intelligent Sensory Evaluation – Methodologies and Applications, pp. 312–326. Springer, Heidelberg (2004)
12. Ruan, D.: Modern approaches and advanced applications for plant surveillance and diagnostics: an overview. In: Ruan, D., Fantoni, P.F. (eds.) Power Plant Surveillance and Diagnostics—Applied Research with Artificial Intelligence, pp. 1–30. Springer, Heidelberg (2002)
13. Zhao, Y., Shi, J., Ruan, D., Shi, P.: Interesting rough lattice-based implication rules discovery. In: Ruan, D., Kacprzyk, J., Fedrizzi, M. (eds.) Soft Computing for Risk Evaluation and Management, pp. 121–132. Physica-Verlag, Heidelberg (2001)

14. Zhao, Y., Ruan, D., Shi, P.: Concept lattice-based approach for incrementally association mining. In: Ruan, D., Kacprzyk, J., Fedrizzi, M. (eds.) Soft Computing for Risk Evaluation and Management, pp. 133–139. Physica- Verlag, Heidelberg (2001)
15. Ruan, D.: The progressing of fuzzy-logic control applications to the Belgian reactor. In: Bouchon-Meunier, B., Yager, R.R., Zadeh, L.A. (eds.) Uncertainty in Intelligent and Information Systems, pp. 262–273. World Scientific, Singapore (2000)
16. Liu, J., Ruan, D., Song, Z., Xu, Y.: On optimization with fuzzy constraint conditions. In: Hampel, R., Wagenknecht, M., Chaker, N. (eds.) Fuzzy Control, Theory and Practice, pp. 166–173. Physica-Verlag, Heidelberg (2000)
17. Xu, Y., Ruan, D., Liu, J.: Approximate reasoning based on lattice-valued propositional logic $L_v pl$. In: Ruan, D., Kerre, E.E. (eds.) Fuzzy If-Then Rules in Computational Intelligence: Theory and Applications, pp. 81–105. Kluwer Academic Publishers, Boston (2000)
18. Zhou, C., Ruan, D.: Fuzzy rules extraction-based linguistic and numerical heterogeneous data fusion for intelligent robotic control. In: Ruan, D., Kerre, E.E. (eds.) Fuzzy If-Then Rules in Computational Intelligence: Theory and Applications, pp. 243–265. Kluwer Academic Publishers, Boston (2000)
19. Ruan, D., Li, X., Van den Eynde, G.: Adaptive fuzzy control for a simulation of hydraulic analogy of a nuclear reactor. In: Ruan, D. (ed.) Fuzzy Systems and Soft Computing in Nuclear Engineering, pp. 65–82. Physica-Verlag, Heidelberg (1999)
20. Ruan, D.: Safety evaluation of nuclear transmission lines using fuzzy relations. In: Mastorakis, N.E. (ed.) Computers and Computational Engineering in Control, pp. 188–192. World Scientific Engineering Society Press (1999)
21. Ruan, D.: Fuzzy logic and intelligent computing in nuclear engineering. In: Zadeh, L.A., Kacprzyk, J. (eds.) Computing with Words in Information/Intelligent Systems, vol. 2, pp. 567–585. Physica-Verlag, Heidelberg (1999)
22. Huang, C., Ruan, D.: System analyse models for fuzzy risk estimation. In: Zadeh, L.A., Kacprzyk, J. (eds.) Computing with Words in Information/Intelligent Systems, vol. 2, pp. 195–220. Physica-Verlag, Heidelberg (1999)
23. Li, X., Ruan, D.: Novel neural algorithms for solving fuzzy relation equations. In: Ruan, D. (ed.) Intelligent Hybrid Systems: Fuzzy Logic, Neural Networks, and Genetic Algorithms, pp. 59–89. Kluwer Academic Publishers, Boston (1997)
24. Ruan, D.: Fuzzy logic applications in nuclear industry. In: Ruan, D. (ed.) Fuzzy Logic Foundations and Industrial Applications, pp. 313–327. Kluwer Academic Publishers, Boston (1996)
25. Huang, C., Ruan, D.: Information diffusion principle and application in fuzzy neuron. In: Ruan, D. (ed.) Fuzzy Logic Foundations and Industrial Applications, pp. 165–198. Kluwer Academic Publishers, Boston (1996)
26. Ruan, D.: Fuzzy decision-making applications in nuclear science. In: Bien, Z., Min, K.C. (eds.) Fuzzy Logic and Its Applications to Engineering, Information Sciences, and Intelligent Systems, pp. 275–284. Kluwer Academic Publishers, Dordrecht (1995)
27. Wang, X., Ruan, D.: On transitivity of fuzzy preference relations in ranking fuzzy numbers. In: Ruan, D. (ed.) Fuzzy Set Theory and Advanced Mathematical Applications, pp. 155–173. Kluwer Academic Publishers, Boston (1995)
28. Ruan, D.: Critical study of fuzzy implication operators and their influence on approximate reasoning. In: Kerre, E.E. (ed.) Introduction to the Basic Principles of Fuzzy Set Theory and Some of Its Applications, pp. 214–251. Communication and Cognition, Gent (1991)

2.4.4 Conference Papers

1. Ruan, D., Hardeman, F., Mkrtchyan, L.: Using belief degree-distributed fuzzy cognitive maps in nuclear safety culture assessment. NAFIPS 2011, March 18-20, El Paso, Texas, USA, pp. 1–6(2011)
2. Mkrtchyan, L., Ruan, D.: Belief degree-distributed fuzzy cognitive maps. In: 2010 IEEE International Conference on Intelligent Systems and Knowledge Engineering, Hangzhou, China, November 15-16, pp. 159–165 (2010)
3. Bao, Q., Ruan, D., Shen, Y., Hermans, E.: Creating a composite road safety performance index by a hierarchical fuzzy TOPSIS approach. In: 2010 IEEE International Conference on Intelligent Systems and Knowledge Engineering, Hangzhou, China, November 15-16, pp. 458–463 (2010)
4. Zhai, Y., Yang, B., Ma, N., Ruan, D.: New construction of ensemble classifiers for imbalanced datasets. In: 2010 IEEE International Conference on Intelligent Systems and Knowledge Engineering, Hangzhou, China, November 15-16, pp. 228–233 (2010)
5. Liu, J., Xu, Y., Ruan, D.: An axiomatizable lattice-ordered linguistic truth-valued logic. In: Ruan, D., Li, T.R., Kerre, E.E., Chen, G.Q., Xu, Y. (eds.) Computational Intelligence – Foundations and Applications, FLINS 2010. World Scientific Proceedings Series on Computer Engineering and Information Science, Emei, Chengdu, China, August 2-4, vol. 4, pp. 72–78 (2010)
6. Buyukozkan, G., Cifci, G., Ruan, D.: An integration of fuzzy analytic network process and fuzzy decision making for mass customization strategies. In: Ruan, D., Li, T.R., Kerre, E.E., Chen, G.Q., Xu, Y. (eds.) Computational Intelligence – Foundations and Applications, FLINS 2010. World Scientific Proceedings Series on Computer Engineering and Information Science, Emei, Chengdu, China, August 2-4, vol. 4, pp. 229–235 (2010)
7. Purba, J.H., Lu, J., Ruan, D., Zhang, G.: Probabilistic safety assessment in nuclear power plants by fuzzy numbers. In: Ruan, D., Li, T.R., Kerre, E.E., Chen, G.Q., Xu, Y. (eds.) Computational Intelligence – Foundations and Applications, FLINS 2010. World Scientific Proceedings Series on Computer Engineering and Information Science, Emei, Chengdu, China, August 2-4, 2010, vol. 4, pp. 256–262. World Scientific, Singapore (2010)
8. Kabak, O., Ruan, D.: Analyzing environmental samples in nuclear safeguards evaluation by a cumulative belief degrees approach. In: Ruan, D., Li, T.R., Kerre, E.E., Chen, G.Q., Xu, Y. (eds.) Computational Intelligence – Foundations and Applications, FLINS 2010, World Scientific Proceedings Series on Computer Engineering and Information Science, Emei, Chengdu, China, August 2-4, vol. 4, pp. 797–805 (2010)
9. Shi, Y., Van Gasse, B., Ruan, D., Kerre, E.E.: A multi-criteria group decision algorithm for policy analysis. In: Ruan, D., Li, T.R., Kerre, E.E., Chen, G.Q., Xu, Y. (eds.) Computational Intelligence – Foundations and Applications, FLINS 2010, World Scientific Proceedings Series on Computer Engineering and Information Science, Emei, Chengdu, China, August 2-4, vol. 4, pp. 832–837 (2010)

10. Shen, Y., Hermans, E., Ruan, D., Vanhoof, K., Brijs, T., Wets, G.: A DEA-based Malmquist productivity index approach in assessing road safety performance. In: Ruan, D., Li, T.R., Kerre, E.E., Chen, G.Q., Xu, Y. (eds.) Computational Intelligence – Foundations and Applications, FLINS 2010, World Scientific Proceedings Series on Computer Engineering and Information Science, Emei, Chengdu, China, August 2-4, vol. 4, pp. 923–928 (2010)
11. Espinilla, M., Ruan, D., Liu, J., Martínez, L.: A heterogeneous evaluation model for assessing sustainable energy: a Belgian case study. In: WCCI 2010 IEEE World Congress on Computational Intelligence, Barcelona, Spain, July 18-23, pp. 227–234 (2010)
12. Purba, J.H., Lu, J., Ruan, D., Zhang, G.: A Hybrid Approach for Fault Tree Analysis Combining Probabilistic Method with Fuzzy Numbers. In: Rutkowski, L., Scherer, R., Tadeusiewicz, R., Zadeh, L.A., Zurada, J.M. (eds.) ICAISC 2010. LNCS, vol. 6113, pp. 194–201. Springer, Heidelberg (2010)
13. Rodríguez, R.M., Ruan, D., Liu, J., Calzada, A., Martínez, L.: Imputing Missing Values in Nuclear Safeguards Evaluation by a 2-Tuple Computational Model. In: Rutkowski, L., Scherer, R., Tadeusiewicz, R., Zadeh, L.A., Zurada, J.M. (eds.) ICAISC 2010. LNCS, vol. 6113, pp. 202–209. Springer, Heidelberg (2010)
14. He, X., Xu, Y., Li, Y., Liu, J., Martinez, L., Ruan, D.: α-Satisfiability and α-Lock Resolution for a Lattice-Valued Logic LP(X). In: Corchado, E., Graña Romay, M., Manhaes Savio, A. (eds.) HAIS 2010. LNCS, vol. 6077, pp. 320–327. Springer, Heidelberg (2010)
15. Pan, X., Xu, Y., Martinez, L., Ruan, D., Liu, J.: On Compactness and Consistency in Finite Lattice-Valued Propositional Logic. In: Corchado, E., Graña Romay, M., Manhaes Savio, A. (eds.) HAIS 2010. LNCS, vol. 6077, pp. 328–334. Springer, Heidelberg (2010)
16. Wang, Y., Ruan, D., Xu, J., Wen, M., Deng, L.: Computational intelligence algorithms analysis for smart grid cyber security. In: Tan, Y., Shi, Y., Tan, K.C. (eds.) ICSI 2010, Part II. LNCS, vol. 6146, pp. 77–84. Springer, Heidelberg (2010)
17. Xu, Y., Liu, J., Ruan, D.: Lattice-valued algebra and logic with potential applications in linguistic valued decision making. In: Proceedings of 31st LINZ Seminar on Fuzzy Set Theory: "Lattice-Valued Logic and Its Applications", Bildungszentrum St. Magdalena, Linz, Austria, February 9-13, pp. 163–167 (2010)
18. Shi, Y., Van Gasse, B., Ruan, D., Kerre, E.E.: A thorough study of interrelationship in fuzzy implication axioms. In: Vanhoof, K. et al (eds.) Intelligent Decision Making Systems, ISKE 2009. World Scientific Proceedings Series on Computer Engineering and Information Science, vol. 2, pp. 52–57 (2009)
19. Kabak, O., Ruan, D.: Dealing with missing values in nuclear safeguards evaluation. In: Vanhoof, K. et al (eds.) Intelligent Decision Making Systems, ISKE2009, World Scientific Proceedings Series on Computer Engineering and Information Science, vol. 2, pp. 145–152 (2009)

20. Liu, J., Martínez, L., Ruan, D., Wang, H.: Generating consistent fuzzy belief rule base from sample data. In: Vanhoof, K., et al. (eds.) Intelligent Decision Making Systems, ISKE 2009. World Scientific Proceedings Series on Computer Engineering and Information Science, vol. 2, pp. 167–172 (2009)
21. Shen, Y., Hermans, E., Ruan, D., Wets, G., Vanhoof, K., Brijs, T.: A generalized multiple data envelopment analysis model. In: Vanhoof, K., et al. (eds.) Intelligent Decision Making Systems, ISKE 2009. World Scientific Proceedings Series on Computer Engineering and Information Science, vol. 2, pp. 215–220 (2009)
22. Buyukozkan, G., Ruan, D.: A combined fuzzy group decision making framework to evaluate agile supply chain enablers. In: Vanhoof, K., et al. (eds.) Intelligent Decision Making Systems, ISKE 2009. World Scientific Proceedings Series on Computer Engineering and Information Science, vol. 2, pp. 299–304 (2009)
23. Kabak, O., Ruan, D.: A cumulative belief-degree approach for nuclear safeguards evaluation. In: Proceedings of the 2009 IEEE International Conference on Systems, Man, and Cybernetics, San Antonio, TX, USA, October 11-14, pp. 2285–2290 (2009)
24. Ma, J., Zhang, G., Lu, J., Ruan, D.: Impute missing assessments by opinion clustering in multi-criteria group decision making problems. In: Proceedings of IFSA/EUSFLAT 2009, Lisbon, Portugal, July 20-24, pp. 555–560 (2009)
25. Li, T., Shen, Y., Ruan, D., Hermans, E., Wets, G.: Integrating Rough Sets with Neural Networks for Weighting Road Safety Performance Indicators. In: Wen, P., Li, Y., Polkowski, L., Yao, Y., Tsumoto, S., Wang, G. (eds.) RSKT 2009. LNCS, vol. 5589, pp. 60–67. Springer, Heidelberg (2009)
26. Ruan, D., Carchon, R., Van der Meer, K., Liu, J., Wang, H., Martínez, L.: Belief rule-based inference methodology to improve nuclear safeguards information evaluation. In: NAFIPS 2009, Cincinnati, Ohio, USA, June 14-17, p. 6 (2009)
27. Hermans, E., Ruan, D., Brijs, T., Wets, G., Vanhoof, K.: Evaluation of road safety performance indicators using OWA operators. In: Ruan, D., Montero, J., Lu, J., Martínez, L., D'hondt, P., Kerre, E.E. (eds.) Computational Intelligence in Decision and Control, The Proceedings of FLINS 2008, Madrid, Spain, September 21-24, pp. 695–700 (2008)
28. Buyukozkan, G., Ruan, D.: A multi-criteria aggregation approach to software development risk management. In: Ruan, D., Montero, J., Lu, J., Martínez, L., D'hondt, P., Kerre, E.E. (eds.) Computational Intelligence in Decision and Control, The Proceedings of FLINS 2008, Madrid, Spain, September 21-24, pp. 555–560 (2008)
29. Song, J., Li, T., Ruan, D.: A new decision tree construction using the cloud transform and rough sets. In: Wang, G., Li, T., Grzymala-Busse, J.W., Miao, D., Skowron, A., Yao, Y. (eds.) RSKT 2008. LNCS (LNAI), vol. 5009, pp. 524–531. Springer, Heidelberg (2008)
30. Ruan, D., Lu, J., Laes, E., Zhang, G., Wu, F., Hardeman, F.: Fuzzy multi-criteria group decision support in long-term options of Belgian energy policy. In: NAFIPS 2007, San Diego, CA, pp. 497–501 (2007)

31. Li, T., Ruan, D., Huang, T., Xu, Y.: On a Mathematical Relationship Between the Fixed Point and the Closed Itemset in Association Rule Mining. In: Zhang, S., Jarvis, R.A. (eds.) AI 2005. LNCS (LNAI), vol. 3809, pp. 989–992. Springer, Heidelberg (2005)
32. Liu, J., Xu, Y., Ruan, D., Martínez, L.: A lattice-valued linguistic-based decision-making method. In: Proceedings of IEEE International Conference on Granular Computing (GrC 2005), Beijing, China, July 25-27, pp. 199–203 (2005)
33. Benitez-Read, J.S., Perez-Cruz, J.H., Ruan, D.: Fusion of state feedback, prediction, and fuzzy logic for power regulation in a research reactor. In: 2005 IEEE International Conference on Systems, Man and Cybernetics, Waikoloa, Hawaii, USA, October 10-12, pp. 3847–3852 (2005)
34. Ruan, D., Maes, D., Aït Abderrahim, H., Liu, J., Xu, L., Yang, J.B.: Cost estimations for large scale engineering projects under various uncertainties. In: Joint EUSFLAT-LFA 2005, CD-ROM Proceedings, Barcelona, Spain, September 7-9, pp. 955–960 (2005)
35. Ruan, D.: Artificial intelligence in nuclear engineering: developments, lessons learned and future directions (invited). In: 2005 International Nuclear Atlantic Conference (INAC), CD-ROM Proceedings, Santos, Sao Paolo, Brazil, August 28-September 2, p. 6 (2005)
36. Benitez-Read, J.S., Perez-Cruz, J.H., Lopez-Callejas, R., Pacheco-Sotelo, J.O., Ruan, D.: Speeding up the power response in a reactor using state feedback and a first order predictive stage. In: Proceedings of the 3rd International Conference on Computing, Communications and Control Technologies, Austin, Texas, USA, July 24-27, vol. 1, pp. 136–141 (2005)
37. Ruan, D.: Applied AI in complex systems: examples and lessons learned (invited). In: The 8th Joint Conference on Information Sciences, CD-ROM Proceedings, Salt Lake City, Utah, USA, July 21-26 (2005)
38. Li, W., Ding, Y., Ruan, D., Huang, Z., Shao, S.: A programmable DNA automation model for context-free grammars. In: The 8th Joint Conference on Information Sciences, CD-ROM Proceedings, Salt Lake City, Utah, USA, July 21-26 (2005)
39. Li, X., Tang, B., Ruan, D.: On three thinking forms: abstract, imagination and intuition based on fuzzy models. In: The 8th Joint Conference on Information Sciences, CD-ROM Proceedings, Salt Lake City, Utah, USA, July 21-26, pp. 1518–1521 (2005)
40. Pei, Z., Ruan, D., Xu, Y.: Gathering linguistic information in distributed intelligent agent on the internet. In: The 8th Joint Conference on Information Sciences, CD-ROM Proceedings, Salt Lake City, Utah, USA, July 21-26, pp. 1569–1572 (2005)
41. Buyokozkan, G., Ruan, D.: Evaluation of software development projects by a fuzzy multi-criteria decision approach. In: The 17th IMACS World Congress, Paris, France, July 11-13, p. 7 (2005)
42. Benítez-Read, J.S., Ruan, D., Ruiz-Enciso, J.A., López-Callejas, R., Pacheco-Sotelo, J.O.: Use of ANN in a Research Reactor Power Fuzzy Controller. In: Cabestany, J., Prieto, A.G., Sandoval, F. (eds.) IWANN 2005. LNCS, vol. 3512, pp. 1132–1139. Springer, Heidelberg (2005)

43. Ruan, D., Benitez-Read, J.S.: Artificial intelligence in nuclear reactor operation. In: The 13th International Conference on Nuclear Engineering, CD-ROM Proceedings, May 16-20, p. 8 (2005)
44. Lu, J., Shi, C., Zhang, G., Ruan, D.: Multi-follower linear bilevel programming: model and Kuhn-Tucker approach. In: Proceedings of the IADIS International "Conference Applied Computing 2005", Algarve, Portugal, February 22-25, vol. 1, pp. 81–88 (2005)
45. Ruan, D., Maes, D., Aït Abderrahim, H., Liu, J., Yang, J.B.: Cost estimations for large scale engineering projects under various uncertainties –case study for the Myrrha machine with intelligent decision systems. In: The IAEA Technical Meeting on "Increasing Instrument Calibration Interval Through on-line Monitoring Technologies", CD-ROM Proceedings, Halden, Norway, September 27-29, p. 12 (2004)
46. Ruan, D.: Fuzzy techniques in control and decision making for complex systems (invited plenary lecture). In: Proceedings of the 12th Spanish Conference on Fuzzy Logic and Application, Jaén, Spain, September 17-19, pp. 9–14 (2004)
47. Xu, Y., Ruan, D., Liu, J.: Progress and prospect in lattice-valued logic systems based on lattice implication algebras. In: Ruan, D., D'hondt, P., De Cock, M., Nachtegael, M., Kerre, E.E. (eds.) Applied Computational Intelligence–Proceedings of the 6th International FLINS Conference, Blankenberge, Belgium, September 1-3, pp. 28–34 (2004)
48. Ma, J., Chen, G., Kerre, E.E., Ruan, D.: A method for association rules mining. In: Ruan, D., D'hondt, P., De Cock, M., Nachtegael, M., Kerre, E.E. (eds.) Applied Computational Intelligence–Proceedings of the 6th International FLINS Conference, Blankenberge, Belgium, September 1-3, pp. 173–178 (2004)
49. Li, T., Ruan, D.: An extended process model of knowledge discovery in database. In: Ruan, D., D'hondt, P., De Cock, M., Nachtegael, M., Kerre, E.E. (eds.) Applied Computational Intelligence–Proceedings of the 6th International FLINS Conference, Blankenberge, Belgium, September 1-3, pp. 185–188 (2004)
50. Martínez, L., Liu, J., Ruan, D.: Fuzzy models to deal with heterogeneous information in decision making problems in engineering processes. In: Ruan, D., D'hondt, P., De Cock, M., Nachtegael, M., Kerre, E.E. (eds.) Applied Computational Intelligence–Proceedings of the 6th International FLINS Conference, Blankenberge, Belgium, September 1-3, pp. 266–271 (2004)
51. Ma, J., Ruan, D., Xu, Y.: A model for semantics of linguistic information with incomparability. In: Ruan, D., D'hondt, P., De Cock, M., Nachtegael, M., Kerre, E.E. (eds.) Applied Computational Intelligence–Proceedings of the 6th International FLINS Conference, Blankenberge, Belgium, September 1-3, pp. 328–333 (2004)
52. Kahraman, C., Buyukozkan, G., Ruan, D.: A fuzzy heuristic multi-attribute conjunctive approach for ERP software selection. In: Ruan, D., D'hondt, P., De Cock, M., Nachtegael, M., Kerre, E.E. (eds.) Applied Computational Intelligence–Proceedings of the 6th International FLINS Conference, Blankenberge, Belgium, September 1-3, pp. 519–524 (2004)

53. Ruan, D.: Nuclear reactor control with fuzzy logic approaches—strengths, weakness, opportunities, and threats. In: Proceedings of the 31st Enlarged Halden Programme Group Meeting, Sandefjord, Norway, May 9-14, vol. C 5.6, p. 12. The volume of the Man-Technology-Organisation (2004)
54. Ruan, D.: Computational intelligence for applied research. In: Proceedings of the 7th JCIS, Cary, NC, USA, September 26-30, pp. 43–46 (2003)
55. Ruan, D.: Lessons learned from computational intelligence in nuclear applications. In: Proceedings of EUSFLAT 2003, Zittau, Germany, September 10-12, pp. 799–802 (2003)
56. Ruan, D.: On soft and hard computing techniques for the operation of nuclear power plants. In: Proceedings of International Conference on Fuzzy Information Processing: Theories and Applications, Beijing, China, March 1-4 (2003)
57. Ruan, D.: Towards fusion of hard and soft computing techniques for the operation of nuclear power plants. In: Proceedings (CD-ROM) of ENC 2002–European Nuclear Conference, Lille, France, October 7-9 (2002)
58. Beskese, A., Kahraman, C., Ruan, D.: Flexibility quantification in computer integrated manufacturing systems based on fuzzy cash flow analysis. In: Ruan, D., D'hondt, P., Kerre, E.E. (eds.) Computational Intelligent Systems for Applied Research–Proceedings of the 5th International FLINS Conference, Gent, Belgium, September 16-18, pp. 171–178 (2002)
59. Liu, J., Ruan, D., Carchon, R.: Synthesis and evaluation analysis of the physical model indicator information by computing with words. In: Ruan, D., D'hondt, P., Kerre, E.E. (eds.) Computational Intelligent Systems for Applied Research–Proceedings of the 5th International FLINS Conference, Gent, Belgium, September 16-18, pp. 511–519 (2002)
60. Roverso, D., Ruan, D., Fantoni, P.F.: Improving feedwater cross-correlation flow measurements in nuclear power plants with artificial neural networks. In: Ruan, D., D'hondt, P., Kerre, E.E. (eds.) Computational Intelligent Systems for Applied Research–Proceedings of the 5th International FLINS Conference, Gent, Belgium, September 16-18, pp. 572–579 (2002)
61. Ruan, D., et al.: Integrating cross-correlation techniques and neural networks for feedwater flow measurement. In: Proceedings (CD-ROM) of SMORN VIII–A Symposium on Nuclear Reactor Surveillance and Diagnostics, Göteborg, Sweden, May 27-31 (2002)
62. Ruan, D., Roverso, D., Fantoni, P.F.: Computational intelligence approaches for parametric estimation and feature extraction of power spectral density. In: Proceedings of EUSEFLAT 2001, An International Conference in Fuzzy Logic and Technology, Leicester, UK, September 5-7, pp. 320–323 (2001)
63. Ruan, D.: Safe experiments on fuzzy control for nuclear reactor operations (invited). In: Proceedings (CD-ROM) of Nuclear Plant Instrumentation, Control and Human-Machine Interface Technologies NPIC&HMIT 2000, Washington, DC, USA, November 13-16 (2000)

64. Xu, Y., Kerre, E.E., Ruan, D., Song, Z.: A study on fuzzy reasoning mechanism based on extension principle. In: Proceedings (CD-ROM) of the 9th IEEE International Conference on Fuzzy Systems, San Antonio, Texas, USA, May 7-10, pp. 185–190 (2000)
65. Ruan, D., Liu, J., Carchon, R.: A lattice-valued linguistic decision model for nuclear safeguards applications. In: Proceedings of the 8th Zittau Fuzzy Colloquium, Zittau, Germany, September 6-8, pp. 284–292 (2000)
66. Liu, J., Xu, Y., Ruan, D.: α-Automated reasoning method based on lattice valued propositional logic LP(X). In: Proceedings of the 8th Zittau Fuzzy Colloguium, Zittau, Germany, September 6 - 8, pp. 60–67 (2000)
67. Kahraman, C., Ruan, D., Tolga, E.: Discounted possibilistic cash flows versus probabilistic cash flows. In: Ruan, D., Aït Abderrahim, H., D'hondt, P., Kerre, E.E. (eds.) Intelligent Techniques and Soft Computing in Nuclear Science and Engineering–Proceedings of the 4th International FLINS Conference, Bruges, Belgium, August 28-30, pp. 214–223 (2000)
68. Liu, J., Ruan, D.: Lattice-valued linguistic expression and its application to the safety evaluation for transmission lines. In: Ruan, D., Aït Abderrahim, H., D'hondt, P., Kerre, E.E. (eds.) Intelligent Techniques and Soft Computing in Nuclear Science and Engineering–Proceedings of the 4th International FLINS Conference, Bruges, Belgium, August 28-30, pp. 389–396 (2000)
69. Ruan, D.: Safety regulations and fuzzy logic control to nuclear reactor. In: 1999 Eusflat-Estylf Joint Conference Proceedings, Palma, Spain, September 22-25, pp. 59–62 (1999)
70. Ruan, D., Van den Eynde, G.: Computer simulation of hydraulic analogy of a nuclear reactor with adaptive fuzzy control. In: Proceedings of International Workshop on Information and Communication Technology for Teaching and Training, Gent, Belgium, September 9-10, pp. 61–73 (1999)
71. Xu, Y., Ruan, D., Liu, J.: Uncertainty automated reasoning in intelligent learning of soft knowledge. In: Proceedings of International Workshop on Information and Communication Technology for Teaching and Training, Gent, Belgium, September 9-10, pp. 29–45 (1999)
72. Ruan, D.: Soft computing for nuclear engineering. In: Proceedings of the 7th Zittau Fuzzy Colloquium, Zittau, Germany, September 8-10, pp. 55–62 (1999)
73. Ruan, D.: On-line experiments of controlling nuclear reactor power with fuzzy logic. In: 1999 IEEE International Fuzzy Systems Conference Proceedings, Seoul, Korea, August 22-25, vol. 3, pp. 1712–1717 (1999)
74. Ruan, D.: Fuzzy logic and soft computing in nuclear engineering. In: Proceedings of the Eighth International Fuzzy Systems Association World Congress, Taipei, Taiwan, August 17-20, vol. 1, pp. 325–329 (1999)
75. Ruan, D., Van den Eynde, G.: Simulating an hydraulic analogy of a nuclear reactor and its adaptive fuzzy logic control. In: Proceedings of ESM 1999, Warsaw, Poland, June 1-4, vol. II, pp. 311–315 (1999)
76. Van der Wal, A.J., Ruan, D.: A plea for a proper focus in robotics research. In: Proceedings of JCIS 1998, RTP, N.C., USA, October 24-28, vol. 1, pp. 76–79 (1998)

77. Li, X., Ruan, D.: An adaptive function for the fuzzy controller of the BR1 reactor. In: Proceedings of the 6th Zittau Fuzzy-Colloquium, Zittau, Germany, September 3-4, pp. 91–97 (1998)
78. Huang, C., Ruan, D.: Advances in information diffusion research. In: Proceedings of FUZZ-IEEE 1998, Anchorage, Alaska, USA, May 4-9, pp. 813–818 (1998)
79. Kerre, E.E., Ruan, D.: The power of fuzzy relations for the representation of imprecise knowledge. In: Proceedings of WAC 1998, Anchorage, Alaska, USA, May 10–14 (1998)
80. Zhou, C., Ruan, D.: Integration of linguistic and numerical information for biped locomotion. In: Ruan, D., Aït Abderrahim, H., D'hondt, P., Kerre, E.E. (eds.) Proceedings of the 3rd International FLINS Workshop on Fuzzy Logic and Intelligent Technologies for Nuclear Science and Industry, Antwerp, Belgium, September 14-16, pp. 294–301 (1998)
81. Li, X., Ruan, D.: Comparative study of fuzzy control, PID control, and advanced fuzzy control for simulating a nuclear reactor operation. In: Ruan, D., Aït Abderrahim, H., D'hondt, P., Kerre, E.E. (eds.) Proceedings of the 3rd International FLINS Workshop on Fuzzy Logic and Intelligent Technologies for Nuclear Science and Industry, Antwerp, Belgium, September 14-16, pp. 424–434 (1998)
82. Ruan, D., Li, X.: R&D on fuzzy control applications to the BR1 research reactor. In: Proceedings of IPMU 1998, Paris, France, vol. 1, pp. 452–459 (1997)
83. Li, X., Ruan, D.: A novel neural algorithm and its application in solving fuzzy relation equations. In: Proceedings of 1997 Annual Meeting of the North American Fuzzy Information Processing Society, Syracuse, New York, September 21-24, pp. 356–361 (1997)
84. Li, X., Ruan, D.: Constructing a fuzzy logic control demo model at SCK·CEN. In: Proceedings of EUFIT 1997, Aachen, Germany, vol. 2, pp. 1408–1412 (1997)
85. Ruan, D., Li, X.: The test of fuzzy logic control with a closed loop at the BR1 reactor. In: Proceedings of the Seventh International Fuzzy Systems Association World Congress (IFSA 1997), Prague, Czech Republic, June 24-29, vol. 4, pp. 126–131 (1997)
86. Ruan, D., Li, X.: Fuzzy logic control applications at SCK·CEN: from a nuclear reactor to a demo model. In: Proceedings of the Joint Conference of Information Sciences (JCIS 1997), North Carolina, USA, March 1-5, vol. 1, pp. 132–135 (1997)
87. Ruan, D.: Fuzzy logic applications in nuclear research areas (invited lecture). In: Proceedings of the Workshop on Science and Modern Technology for Safeguards, Arona, Italy, October 28-31, pp. 203–212 (1996)
88. Ruan, D.: The BR1 research facilities to calibrate fuzzy logic technology for nuclear reactor control. In: Aït Abderrahim, H. (ed.) Research Facilities for the Future of Nuclear Energy, pp. 372–381. World Scientific (1996)
89. Resconi, G., Van der Wal, A.J., Ruan, D.: Application of the concept of local equilibrium to solve the Fokker-Planck equation. In: Ruan, D., D'hondt, P., Govaerts, P., Kerre, E.E. (eds.) Proceedings of the 2nd International FLINS Workshop on Intelligent Systems and Soft Computing for Nuclear Science and Industry, Mol, Belgium, September 25-27, pp. 82–88 (1996)

90. Li, X., Ruan, D., Huang, C.: An extended fuzzy neural network. In: Ruan, D., D'hondt, P., Govaerts, P., Kerre, E.E. (eds.) Proceedings of the 2nd International FLINS Workshop, Intelligent Systems and Soft Computing for Nuclear Science and Industry, Mol, Belgium, September 25-27, pp. 145–152 (1996)
91. Yu, H., Zou, K., Ruan, D.: Construction of classification function in diesel engine fault diagnosis. In: Ruan, D., D'hondt, P., Govaerts, P., Kerre, E.E. (eds.) Proceedings of the 2nd International FLINS Workshop, Intelligent Systems and Soft Computing for Nuclear Science and Industry, Mol, Belgium, September 25-27, pp. 270–275 (1996)
92. Liu, X., Ruan, D.: Application of fuzzy decision making to countermeasure strategies after a nuclear accident. In: Ruan, D., D'hondt, P., Govaerts, P., Kerre, E.E. (eds.) Proceedings of the 2nd International FLINS Workshop ,Intelligent Systems and Soft Computing for Nuclear Science and Industry, Mol, Belgium, September 25-27, pp. 308–315 (1996)
93. Liu, Z., Ruan, D.: Experiments of fuzzy logic control on a nuclear research reactor. In: Ruan, D., D'hondt, P., Govaerts, P., Kerre, E.E. (eds.) Proceedings of the 2nd International FLINS Workshop, Intelligent Systems and Soft Computing for Nuclear Science and Industry, Mol, Belgium, September 25-27, pp. 336–348 (1996)
94. Ruan, D.: R&D on fuzzy logic applications at SCK·CEN (invited). In: Proceedings of NAFIP 1996, Berkeley, California, June 19-22, pp. 428–432 (1996)
95. Ruan, D.: Soft and intelligent computing for nuclear industry. In: Proceedings of International Panel Con. on Soft and Intelligent Computing, Budapest, Hungary, September 30-October 3, pp. 253–258 (1996)
96. Ruan, D., Yin, X., Van De Durpel, L., D'hondt, P.: Application of fuzzy control for nuclear research reactor. In: Proceedings of the 4th IFSICC-International Fuzzy Systems and Intelligent Control Conference, Maui, Hawaii, USA, April 8-11 (1996)
97. Ruan, D.: Application of fuzzy logic to nuclear industry (invited). In: Proceedings of Symposium on Knowledge and Information Technology, Gent, Belgium, September 23-24, pp. 131–140 (1996)
98. Ruan, D., Liu, Z., Van Den Durpel, L., D'hondt, P.: Progress of fuzzy logic control applications for the Belgian Nuclear Reactor 1. In: Proceedings of EUFIT 1996, vol. 2, pp. 1520–1524 (1996)
99. Ruan, D., Kerre, E.E.: On If-Then-Else inference rules. In: Proceedings of 1996 IEEE International Conference on SMC 1996, Beijing, China, October 14-17, pp. 1420–1425 (1996)
100. Yin, X., Ruan, D., Van Den Durpel, L., D'hondt, P.: Development of a fuzzy logic controller at the Belgian reactor 1. In: Proceedings of EUFIT 1995, Aachen, Germany, August 28-31 (1995)
101. Ruan, D.: FLINS: a bridge between fuzzy logic and the nuclear industrial world. In: Proceedings of ISFL 1995, International Symposium on Fuzzy Logic, Zurich, Switzerland, May 26-27 (1995)
102. Ruan, D.: Information & knowledge engineering in nuclear applications (invited plenary lecture). In: Proceedings of International Conference on Information & Knowledge Engineering (ICIK 1995), Dalian, China, August 21-25, pp. 46–60 (1995)

103. Wang, X., Ruan, D., Kerre, E.E.: The use of weak transitivity in ranking fuzzy numbers. In: Proceedings of International Conference on Information & Knowledge Engineering (ICIK 1995), Dalian, China, August 21-25, pp. 322–332 (1995)
104. Ruan, D., Van Den Durpel, L., D'hondt, P.: Fuzzy engineering in nuclear research applications. In: Proceedings of FUZZ-IEEE/IFES 1995, Yokohama, Japan, March 20-24, pp. 211–218 (1995)
105. Ruan, D., Kerre, E.E.: Extended fuzzy chaining syllogism. In: Proceedings of IFSA 1995, São Paulo, Brazil, July 22-28, pp. 145–148 (1995)
106. Ruan, D., De Raedt, C., D'hondt, P., Mund, E.: TRANSFUSION: A time-dependent finite-element code for the solution of the diffusion equation in nuclear oil well logging applications. In: Proceedings of the 10th International Conference on Mathematical and Computer Modelling and Scientific Computing, Boston, Massachusetts, USA, July 5-8 (1995)
107. Wang, X., Ruan, D.: On transitivity properties and ordering of alternatives in fuzzy decision making. In: Proceedings of the 10th International Conference on Mathematical and Computer Modelling and Scientific Computing, Boston, Massachusetts, USA, July 5-8 (1995)
108. Van den Durpel, L., Ruan, D.: Comparison study on a period-generated control law and a fuzzy model-based control in the Belgian nuclear reactor 1. In: Proceedings of EUFIT 1994, Aachen, Germany, September 20-23, pp. 1048–1049 (1994)
109. Wang, X., Kerre, E.E., Ruan, D.: Consistency of judgment matrix and fuzzy weights in fuzzy analytic hierarchy process. In: Proceedings of EUFIT 1994, Aachen, Germany, September 20-23, pp. 157–161 (1994)
110. Wang, X., Kerre, E.E., Cappelle, B., Ruan, D.: Transitivity of fuzzy orderings based on pairwise comparisons. In: Ruan, D., D'hondt, P., Govaerts, P., Kerre, E.E. (eds.) Proceedings of the 1st International FLINS Workshop, Fuzzy Logic and Intelligent Technologies in Nuclear Science, Mol, Belgium, September 14-16, pp. 38–43 (1994)
111. Van Den Durpel, L., Ruan, D.: Fuzzy model-based control of a nuclear reactor. In: Ruan, D., D'hondt, P., Govaerts, P., Kerre, E.E. (eds.) Proceedings of the 1st International FLINS Workshop on Fuzzy Logic and Intelligent Technologies in Nuclear Science, Mol, Belgium, September 14-16, pp. 247–253 (1994)
112. Van de Walle, B., Ruan, D., Kerre, E.E.: Implementing the generalized Modus-Ponens in a fuzzy rule-based for an atmospheric stability monitoring study. In: Proceedings of IEEE World Congress on Computational Intelligence, Orlando, Florida, June 26-July 2, pp. 1081–1086 (1994)
113. Van de Walle, B., Van Camp, M., Ruan, D.: Development of a fuzzy monitoring and early-warning control system for nuclear emergencies. In: Proceedings of International Fuzzy Systems and Intelligent Controls Conference, Louisville, Kentucky, USA, March 13-16 (1994)
114. De Raedt, C., Ruan, D.: Present development of TRANSFUSION, a novel time-dependent three-dimensional finite-element neutron and gamma diffusion code for the analysis of pulsed-neutron oil-well logs. In: Proceedings of International Conference on Nuclear Data for Science and Technology, Gatlinburg, Tennessee, May 9-13 (1994)

115. Ruan, D.: Fuzzy sets and decision making in nuclear science. In: Proceedings of the 5th IFSA World Congress, Seoul, Korea, July 4-9, pp. 1058–1061 (1993)
116. Van den Durpel, L., Ruan, D., D'hondt, P.: Use of fuzzy set theory in the inspection of transmission lines of nuclear installations. In: Proceedings of the 5th IFSA World Congress, Seoul, Korea, July 4-9, pp. 1066–1069 (1993)
117. Ruan, D., D'hondt, P.: Fuzzy systems in nuclear applications. In: Proceedings of EUFIT 1993, Aachen, Germany, September 7-10, pp. 87–90 (1993)
118. Van Camp, M., Ruan, D., Sohier, A., Govaerts, P.: The use of fuzzy sets theory to reduce uncertainties on the source term and the wind direction in decision-aiding systems. In: Proceedings of International Conference on Mathematical Methods and Supercomputing in Nuclear Applications, Karlsruhe, Germany, April 19-23 (1993)
119. Ruan, D., De Raedt, C., Zhang, F.: Development work currently being carried out on the time-dependent finite-element diffusion code TRANSFUSION for nuclear oil well logging problems. In: Proceedings of International Conference on Mathematical Methods and Supercomputing in Nuclear Applications, Karlsruhe, Germany, April 19-23, pp. 81–92 (1993)
120. Van de Walle, B., Ruan, D., Govaerts, P.: Atmospheric stability modeling for nuclear emergency response systems using fuzzy set theory. In: Proceedings of ASIA Congress on Radiation Protection, Beijing, China, October 18-22 (1993)
121. Ruan, D.: Fuzzy sets and decision theory (invited plenary lecture). In: Proceedings of Conference on Fuzzy Logic, Antwerp, Belgium, November 25-26, p. 30 (1992)
122. Sohier, A., Van Camp, M., Ruan, D., Govaerts, P.: Methods for radiological assessment in the near-field during the early phase of an accidental release of radioactive material using an incomplete data base. In: Proceedings of the 3rd International Workshop on Real-Time Computing of the Environmental Consequences of an Accidental Release to Atmosphere from a Nuclear Installation, October 25-30, Schloss Elman, Bavaria (1992)
123. Tang, H., Ruan, D.: Land use suitability assessment for irrigated maize based on fuzzy set theory. In: Proceedings of International Conference Advances in Planning, Design and Management of Irrigation Systems as Related to Sustainable Land Use, Leuven, Belgium, September 14-17, pp. 597–605 (1992)
124. De Raedt, C., Ruan, D.: TRANSFUSION: a novel three-dimensional time-dependent finite-element diffusion code for modeling pulsed-neutron logging tool. In: Proceedings of the 4th EC Symposium on Oil and Gas Technology in a Wide Europe, Berlin, Germany, November 3-5 (1992)
125. Yang, G., Ruan, D.: DBCDSP: Downhill belt conveyer dynamical simulation in Pascal. In: Proceedings of the 7th IFAC Symposium on Automation in Mining, Mineral and Metal Processing, Beijing, China, August 26-28, pp. 28–32 (1992)
126. De Cooman, G., Kerre, E.E., Cappelle, B., Ruan, D., Vanmassenhove, F.: On the extension of classical propositional logic by means of a triangular norm. In: Proceedings of the 3rd IFSA Congress, Seattle, USA, pp. 821–824 (1989)
127. Ruan, D., Kerre, E.E., de Cooman, G., Cappelle, B., Vanmassenhove, F.: The influence of the fuzzy implication operator on the method-of-cases inference rule. In: Proceedings of Knowledge-Based Systems and Models of Logical Reasoning, p. 8. Scomi Scientific Center, El-Cherif (1988)

2.4.5 Technical Reports

1. Ruan, D., Roverso, D., Fantoni, P.F., Sanabrias, J.I., Carrasco, J.A., Fernandez, L.: Feedwater flow measurements: challenges, current solutions, and "soft" development, OECD Halden Reactor Project, HWR-692, p. 38 (2002)
2. Ruan, D., Carchon, R., Kerre, E.E.: Aggregation operators: properties and applications, Task performed in the frame of the Belgian support programme to the IAEA for safeguards implementation, task no. C01181, SCK·CEN Report, R-3331, p. 16 (1999)
3. Ruan, D., de Cooman, G., Ryjov, A.P.: Fuzzy Logic and Intelligent Technologies in Russia, BLG-669, p.163p. SCK·CEN, Mol, Belgium (1995)
4. Ruan, D.: Basic Concepts in Nuclear Research – Core Activities at the Belgian Nuclear Research Centre, BLG-653, p. 50. SCK·CEN, Mol, Belgium (1994)
5. Ruan, D.: Intelligent Technologies for Man-Machine Interaction at the OECD Halden Reactor Project - Special Presentation of FLINS, BLG-665, p. 51. SCK·CEN, Mol, Belgium (1994)
6. Ruan, D., De Raedt, C.: Development of TRANSFUSION: a three-dimensional time-dependent finite element diffusion code for logging problems, ABSTRACT Five Years Supercomputing, pp. 49-51. University of Leuven, Belgium (1987-1992)

2.5 Chairing the FLINS Conference Series (1994-2012)

Profs. Drs. Etienne Kerre[1], Martine De Cock[1], Mike Nachtegael[1], Javier Montero[2], Yang Xu[3], and Tianrui Li[3]

[1] Ghent University, Gent, Belgium
[2] Complutense University of Madrid, Madrid, Spain
[3] Southwest Jiaotong University, School of Mathematics, Chengdu, China

Etienne's General Review of FLINS

Prof. Da Ruan took the initiative in the FLINS conferences and has been the chair from the first to the 7th, and the general chair from the 8th to the 10th FLINS conferences.

FLINS, an acronym for Fuzzy Logic and Intelligent Technologies in Nuclear Science, is a well-established international research forum to advance the theory and applications of computational intelligence for applied research in general and for complex engineering and decision support in particular. The principal mission of FLINS is bridging the gap between machine intelligence and real complex systems via joint research between universities and international research institutions, by encouraging interdisciplinary research and bringing multi-discipline researchers together.

- The 1st International FLINS Workshop took place September 14-16, 1994, Mol, Belgium. There were a total of 56 papers selected out of 70 regular submissions and 4 invited plenary papers from 13 countries. Invited keynote speakers were Prof. H. J. Zimmermann from Germany, Prof. E. E. Kerre from Belgium, Prof. G. Klir from USA, and Prof. A. J. van der Wal from the Netherlands.
- FLINS 1996, the 2nd International FLINS Workshop on Intelligent Systems and Soft Computing for Nuclear Science and Industry took place September 25-27, 1996, Mol, Belgium. There were a total of 52 papers selected out of 80 regular and 6 invited plenary papers submissions from 20 countries. Invited keynote speakers were Prof. H. J. Zimmermann from Germany, Prof. E. E. Kerre from Belgium, Prof. P. Smets from Belgium, Profs. M. Modarres, P. P. Wang and X. L. Gu from USA, and Profs. H. Bersini and G. Bontempi from Belgium.

- FLINS 1998, the 3rd International Workshop on Fuzzy Logic and Intelligent Technologies for Nuclear Science and Industry took place September 14-16, 1998, Antwerp, Belgium. There were a total of 55 papers selected out of 80 regular submissions and 7 invited plenary papers from 23 countries. Invited keynote speakers were Prof. L. A. Zadeh from USA, Prof. M. M. Gupta from Canada, Prof. J. Kacprzyk from Poland, Prof. E. E. Kerre from Belgium, Prof. A. J. van der Wal from the Netherlands, Prof. R. R. Yager from USA, and Prof. H. J. Zimmermann from Germany.
- FLINS 2000, the 4th International Conference on Intelligent Technologies and Soft Computing in Nuclear Sciences and Engineering took place August 28-30, 2000, Bruges, Belgium. There were a total of 65 papers selected out of 90 submissions and 6 invited plenary papers from 28 countries. Invited keynote speakers were Prof. L. A. Zadeh from USA, Prof. J. Kacprzyk from Poland, Prof. H. J. Zimmermann from Germany, Prof. P. P. Wang from USA, Prof. M. M. Gupta from Canada and Prof. Z. Z. Bien from South Korea.
- FLINS 2002, the 5th International FLINS Conference on Computational Intelligent Systems for Applied Research took place September 16-18, 2002, Gent, Belgium. There were a total of 70 papers selected out of 95 submissions and 4 invited plenary papers from 30 countries. Invited keynote speakers were Profs. L. A. Zadeh, R. E. Uhrig, P. P. Wang from USA and Prof. H. Prade from France.
- FLINS 2004, the 6th International FLINS Conference on Applied Computational Intelligence took place September 1-3, 2004, Blankenberge, Belgium. There were a total of 113 papers selected out of 162 regular submissions and 8 invited papers from no less than 29 countries. Invited keynote speakers were Prof. H.-J. Zimmermann from Germany, Profs. R. R. Yager and Wang from USA, Prof. Z. Z. Bien from South Korea, Prof. M. M. Gupta from Canada, Prof. J. Montero from Spain, Profs. G. Q. Chen and Y. Xu from China. In order to recognize the invited speakers' contributions to the previous FLINS conferences from the 2004 edition onwards each time several FLINS awards were given. The first FLINS Gold Award was given to Prof. L. Zadeh for his invention of fuzzy set theory and for his continuously supporting the FLINS activities. The 2004 FLINS Appreciation Awards went to Profs. H. J. Zimmermann, R.R Yager, P. Wang, Z. Bien, M. Gupta, J. Montero, GQ. Chen and to P. Govaerts, the general manager of SCK•CEN. The 2004 FLINS Outstanding Service Award went to Prof. Y. Xu for his tremendous efforts to attract many Chinese researchers to FLINS 2004.
- FLINS 2006, the 7th International FLINS Conference on Applied Artificial Intelligence took place August 29-31, 2006, Genova, Italy. There were a total of 131 selected out of 207 regular submission papers and 7 invited plenary papers from 25 countries. Invited keynote speakers were Profs. L. A. Zadeh and P. P. Wang from USA, Prof. Z. Z. Bien from South Korea, Prof. M. M. Gupta from Canada, Prof. D. Mundici from Italy, Prof. F. Owre from Norway and Prof. E. Sanchez from France. The 2006 FLINS Gold Medal was given to

Profs. Z. Bien, D. Mundici, F. Owre, E. Sanchez and M. Marseguerra for their support to the FLINS conferences. The 2006 FLINS Outstanding Service Award went to Prof. C. Kahraman for his tremendous efforts to attract many Turkish researchers to FLINS 2006.

- FLINS 2008, the 8th International FLINS Conference on Computational Intelligence in Decision and Control took place September 21-24, 2008, Madrid, Spain. There were a total of 184 papers selected out of 292 regular submissions and 6 invited plenary papers from 35 countries. Invited keynote speakers were Prof. P. Bonissone from USA, Prof. W. Pedrycz from Canada and Prof. E. Trillas from Spain. During the Gala dinner all session chairs as well as the young local organizers got a 2008 FLINS Pin Award. The 2008 FLINS Silver Pin awards were given to Dr. P. D'Hondt, to Profs. L. Magdalena, A. Murari and B. Vitoriano. The 2008 FLINS Golden Pin Awards were awarded to Profs. P. Bonissone, W. Pedrycz and E. Trillas. Prof. L.A. Zadeh got a special 2008 FLINS Friendship Award.
- FLINS 2010, the 9th International FLINS Conference on Computational Intelligence Foundations and Applications took place August 1-4, 2010, Chengdu, China. There were a total of 177 papers selected out of 263 regular submissions and 4 invited plenary papers from 32 countries. Invited keynote speakers were Prof. Z. Z. Bien from South Korea, Prof. B. Bouchon-Meunier from France, Prof. M. M. Gupta from Canada and Profs. V. Novak and I. Perfilieva from Czech Republic. The 2010 FLINS Silver Pin awards were given to Profs. J. Montero, T. Li and C. Kahraman. Prof. E. Kerre received from a very emotional Prof. Da Ruan the FLINS Golden Chair award for his continuous support to the FLINS conferences since their inception in 1994.
- FLINS 2012, the 10th International FLINS Conference on Uncertainty Modeling in Knowledge Engineering and Decision Making will be taking place August 27-29, 2012, Instanbul, Turkey. Invited keynote speakers will be Prof. A. Maisseu from France, Prof. A. W. Deshpande from India, Prof. J. Lu from Australia, Prof. I. B. Turksen from Turkey and Prof. H. J. Zimmerman from Germany.

All of us expect many successful upcoming editions of FLINS conferences in the future as Prof. Da Ruan would have desired.

Martine and Mike's Memories

We were closely involved by Da in the organisation of 3 FLINS conferences, respectively in Bruges (2002), Ghent (2004) and Genova (2006). We learned to know Da as a very engaged conference organizer who put a lot of effort and energy into his project. With an eye for every detail, he even chose the design and colour for the covers of the FLINS proceedings himself. His aim was for the spines of the books to form a rainbow of colours when placed next to each other on a book shelf.

The FLINS Editions of 2002, 2004 and 2006

Da in FLINS Conference in 2002

At the same time we experienced quite a lot of respect and freedom from Da. He engaged us as "FLINS co-managers" in 2002 and let us deal with nearly all FLINS correspondences, which mainly dealt with all the email traffic on submitted papers. We also got to organize the review process – of course with the help of Da and other FLINS organizers – which finally resulted in the difficult process of selecting the accepted papers. We vividly remember how we struggled with several so-called borderline papers (do we accept this paper or not?) and how Da came to the rescue during one of the most intense periods in our young scientific careers. In the beginning of May 2002 we both defended our PhD (with just one weekend between the public defence), and the hours before Mike's PhD were not marked by silent concentration, but by Da and us browsing through the papers, giving comments and finally making acceptance/rejection decisions based on the reviews. The publisher's deadline was close, and, to get the most out of every valuable minute, Da insisted that we even continued the process while Mike was changing into formal clothes for his PhD defence. The PhD-defence itself – with Da as one of the jury members – was in that sense only an "unavoidable interruption", with the selection of accepted papers continuing after the defence... It is quite a nice scene to look back at!

The FLINS 2002 conference was our first experience in organizing a conference. It was very intense, but we truly appreciate to have been given this chance. Da valued our efforts and in 2004 we got some more responsibilities and made it to co-editors of the proceedings. In 2006 we also got full financial responsibility. We will always be grateful for these experiences.

Da in FLINS Conference in 2004

Javier's Memories

It is extremely difficult to maintain a strictly academic discourse about a colleague that unexpectedly passed away when he was a friend and when he has been playing a key role in your recent research activities. He will be deeply missed. In this short note we shall bring here some highlights of the relationship between our research team and Da Ruan, starting with our relationship with the International FLINS Conferences.

Since the beginning of the contacts of members of this Complutense University team on Decision Aid Systems with Da Ruan and his collaborators at SCK•CEN and the University of Ghent, the FLINS conference appeared for us as a singular conference, where practitioners and theoretical researchers could find an easy environment for interchanging knowledge and developing joint projects. FLINS initiative (acronym standing for **F**uzzy **L**ogic and **I**ntelligent Technologies in **N**uclear **S**cience) was launched by 1994 in line with the Belgian Nuclear Research Centre (SCK•CEN) and the initial objective of giving PhD and Postdoc researchers the opportunity to carry out future-oriented research. Since then, objectives have evolved into much wider interests, covering now any research within computational intelligence and knowledge engineering in decision and control, still keeping the foundational declared objective of producing a real advance in the theory and applications of decision making and control for complex systems.

Da in FLINS Conference in 2006

Since its conception, Prof. Da Ruan put into the organization of each FLINS conference an extremely careful effort to facilitate such a scientific interchange, indeed keeping a kind attention for the needs of all conference attendants to create a good climate for communication. Coming to FLINS conference has been always a wonderful experience, both scientific and personal. Beside such a unique approach for a rather small but still somehow heterogeneous community open to any issue related to complex systems, a number of illuminating talks by colleagues come to our mind. But we have also to remind with pleasure all those scientific discussions we could have with colleagues during FLINS conferences, sometimes around a pot of Belgian mussels, or simply around a beer after the conference, or even walking along a beach or some historical site (needless to say, not taking care too much about the surrounding beauty because of the intense scientific conversation we were having). Such a kind of human relationship between the conference attendants created a quite familiar ambience, which cannot be understood without Da's personality, his continuous smile and the devoted dedication of all his FLINS organization. Particularly, I will be always grateful to Da and the FLINS community for the kindness, support and friendship I found in some difficult personal times. We are proud of being a part of his FLINS project.

The FLINS Edition of 2008

Da Ruan gave this research team at Complutense University of Madrid one of the most important moments: the organization of the FLINS 2008 conference in Madrid. The organization of this conference represents a cornerstone on the history of our research group. We learned a lot from previous organizers, we jumped in international visibility, and we sincerely enjoyed it very much.

Da in FLINS Conference Gala Dinner in 2008

The 8th International FLINS Conference on Computational Intelligence and Decision Control was held in Madrid (Spain), September 21-24, 2008, with the participation of scientists coming from 35 different countries and the presentation of 184 papers, selected in a strict peer review process from the 292 submitted papers. A short version of all accepted papers has been published in the FLINS 2008 proceedings. Moreover, special issues with selected extended papers were accepted after a second strictly peer-reviewed process according to the regular standards of five scientific journals: Information Sciences, the International Journal of Computational Intelligence Systems, the International Journal of General Systems, Mathware & Soft Computing and TOP, the official journal of the Spanish Society of Statistics and Operations Research, The FLINS 2008 special issue in the International Journal of Computational Intelligence Systems, since then the official journal of the European Society for Fuzzy Logic and Technologies (EUSFLAT), represents a nice example of the spirit of FLINS, since it is a direct consequence of an intense discussion produced during a FLINS plenary talk, counting with the participation of P. Bonissone, H. Bustince, S. Guadarrama, J. Kacprzyk, L. Magdalena, L. Maguire, C. Moraga, W. Pedrycz, E. Ruspini, R. Seising, E. Trillas, L.A. Zadeh and Da Ruan himself.

Based upon our previous collaborations and some joint projects, Professor Da Ruan joined our research project TIN2009-07901, funded by the Government of Spain. Da Ruan was currently member of this project, to be finished by Dec 31, 2012.

Yang and Tianrui's Memories

During the FLINS2008 conference in Madrid, all of FLINS delegates were concerned about who will host the next FLINS. It was known that at least three universities, including Southwest Jiaotong University, China, Istanbul University of Technology, Turkey and Federal University of Paraiba, Brazil, applied to host FLINS2010. Especially Prof. Cengiz Kahraman and his team, nearly 20 participants, from Istanbul University of Technology, Turkey attended FLINS2008 and hoped to host FLINS2010. In fact, the decision of the host of FLINS2010 by Prof. Da Ruan et al. was as very difficult. However, they considered that FLINS has been held in Europe eight times and it was the right time to promote FLINS in Asia. They also considered that Southwest Jiaotong University, China had successfully hosted ISKE2007 and had many experiences on the organization of international conferences. Therefore, they finally selected Southwest Jiaotong University, China as the host of FLINS2010. At that time, they also decided to select Istanbul University of Technology, Turkey to host the next FLINS.

The FLINS Edition of 2010

In order to better organize FLINS2010, the first concerned thing by Prof. Da Ruan was the selection of the venue of the conference. Six sites of Chengdu and E'Mei, China were recommended and all the information of them was sent to him. After investigating the materials, he made decision that it is better to hold FLINS2010 at E'Mei. Therefore, he chose the three sites at E'Mei as the candidate venues for FLINS2010. In 2009, he personally inspected the three candidate sites by the occasion to lecture in Chengdu, China. At that time, one of his feet was just not in a good condition. A long walk was very difficult for him. However, he put himself out of the way and still adhered to visit the three candidate sites. Finally, the Hongzhu Shan Hotel, E'Mei was selected as the venue of this conference.

Secondly, Prof. Da Ruan gave meticulous guidance on the organization of the conference. He was also in charge of the most important task to contact the press for the proceedings and editors-in-chief of international journals for the special issues of the conference. In his efforts, several famous journals, including International Journal of General Systems, International Journal of Uncertainty, Fuzziness and Knowledge-based Systems, Knowledge-based Systems, Journal of Multiple-Valued Logic and Soft Computing, International Journal of Nuclear Knowledge Management, etc agreed to publish special issues of FLINS2010. During the preparatory of the conference, he worked hard to revise the notice of Call for papers, edit the proceedings, and design the program. Particularly, he found a symbolic error in one of papers of FLINS2010 during the proofreading process of the proceedings. He thought we should contact the authors as soon as possible to confirm the changes due to the limit time for publishing. We remember at that day we had a class at the new campus of our university. Although the time in China was close to the noon, it was in the early morning in Belgium. He had already started to work and called me several times. Since we could not answer the phone during the class, he finally reached me after we finished the class. Later, we promptly contacted the authors and successfully completed the proofreading work. The proceedings of FLINS2010 were then published in time.

Da delivered a welcome speech at the opening ceremony of FLINS2010

In addition, before the start of the FLINS2010 conference, Prof. Da Ruan specifically took aside his time and came to Chengdu a little early to help us prepare the opening and closing ceremonies of the conference. He ignored the time difference and immediately joined us on the preparation of the conference. Specifically, he took his rest time to revise each word of my speeches at opening and closing ceremonies of FLINS2010. Moreover, in the arrangement of the banquet, we did not consider the awards and performance session during the course of the banquet. He reminded us timely and joined us to amend the agenda that very night. Ten activities including awards and performances were then scheduled during the banquet. Through these activities, the whole atmosphere of the banquet was more exciting. Delegates interacted actively. The unexpected awards and perfect performances again and again made the atmosphere to reach its climax. In particular, delegates were deeply impressed by the improvisation of Spanish flamenco dance. More importantly, these activities made delegates not only enjoy the academic "feast", but also feel the joy of life and have a lot of unforgettable memories at FLINS2010.

In the closing ceremony, Prof. Da Ruan announced that the tenth FLINS conference will be hosted by Istanbul University of Technology, Turkey. It drew FLINS2010 to a successful close. Then, delegates gave a warm applause to Prof. Cengiz Kahraman and his team, congratulated them and acknowledged them for their tireless efforts to support FLINS. It is also reflected the FLINS conference initiated by Prof. Da Ruan has a great international influence.

2.6 Chairing the ISKE Conference Series (2006-2011)

Profs. Drs. Yang Xu and Tianrui Li

Southwest Jiaotong University, Chengdu, China

To promote international cooperation of the research on intelligent systems and knowledge engineering, Prof Da Ruan, together with Profs Etienne E. Kerre (Belgium), Guoqing Chen (China), Yang Xu (China), Geert Wets (Belgium), Jie Lu (Australia), Yongsheng Ding (China) originated the 1st International Conference on Intelligent Systems and Knowledge Engineering (ISKE). It was based on the 4th bilateral project of international cooperation between China and Flanders which title is "Intelligent Systems for Data Mining and Information Processing: Methods and Applications". It aims to cover the-state-of-the-art research and development in all aspects related to intelligent systems for knowledge engineering. The 1st ISKE conference was successfully held in Shanghai at April, 2006 and hosted by Donghua University, China. There were nearly 200 researchers, professors, and students from 12 countries and regions participating in ISKE2006 and communicating through a total of 16 regular sessions, 3 keynote lectures and 2 featured lectures.

Prof. Da Ruan at the banquet of ISKE2006, Shanghai, China

Now ISKE series conference has been evolved as an annual international conference. The current conference of ISKE2011 (Shanghai, China, December, 2011) is the 6th in a series of conferences on Intelligent Systems and Knowledge Engineering, following the successful 1st ISKE conference in Shanghai April 2006, 2nd conference in Chengdu, China, October 2007, 3rd conference in Xiamen, China, October 2008, 4th conference in Hasselt, Belgium, November 2009, 5th conference in Hangzhou, China, November 2010. In the meantime, the 7th ISKE conference has been planned in Beijing, China in 2012. Lots of researchers have participated in the academic event. Table 1 shows the statistics of submissions and accepted papers of ISKE conferences from 2006 to 2011.

During the process of preparing every ISKE conference, Prof Da Ruan helped a lot to make ISKE conference a success. He not only assisted on the preparation of the conference, *e.g.*, the notice for call for papers and invitation of PC members, but also took care of selection of accepted papers and the program of every ISKE conference. In addition, he worked hard with the local organization of ISKE conference on the issues of proceedings' publishing. Under his effort, the proceedings of the ISKE conferences were successfully published by IEEE press, World Scientific, Springer and Atlantis press. Papers in the proceedings of the past ISKE conferences have been indexed by EI or ISTP.

Table 1. The statistics of ISKE conferences

ISKE	2006	2007	2008	2009	2010	2011
Submissions (Countries and regions)	198 (14)	797 (19)	549 (22)	152 (26)	317 (22)	605 (26)
Accepted papers (Countries and regions)	119 (12)	284 (16)	270 (19)	108 (22)	134 (19)	261 (23)

Prof Da Ruan carefully selected the keynote speakers for international series of conferences on ISKE. He tried his best to make the ISKE conference more attractive by researchers. Many distinguished professors, e.g., Profs. Lotfi A. Zadeh (USA), Bo Zhang (China), Ruqian Lu (China), Ronald R. Yager (USA), Hans Juergen Zimmermann (Germany), Witold Pedrycz (Canada), Bernadette Bouchon-Meunier (France), Qiang Yang (China), Hugo de Garis (China), Jie Lu (Australia), Fuchun Sun (China), Zhi-Hua Zhou (China), etc, have been invited by Prof. Da Ruan to deliver keynotes in ISKE conferences. He was very happy that their excellent reports benefited a lot for participants of ISKE conference from around the world.

These international series of conferences on ISKE were supported by many universities and institutes in the world, e.g., Southwest Jiaotong University, China, Belgian Nuclear Research Centre, Belgium, Tsinghua University, China, Donghua University, China, Xiamen University, China, Zhejiang University, China, Shanghai Jiao Tong University, China, Ghent University, Belgium, Hasselt University, Belgium, University of Technology, Sydney, Australia, and California State University, USA. The international journals, including Knowledge-based Systems, Soft Computing, International Journal of Computational Intelligence Systems, have published the special issues of ISKE conference, respectively. This is credited to the continuous efforts of Prof. Da Ruan.

Prof. Da Ruan delivered a speech at ISKE2007, Chengdu, China

Prof. Da Ruan delivered a welcome speech at ISKE2010, Hangzhou, China

However, he told his friends that "Though the ISKE conference had become one of important international conferences and had a significant influence on the research of future intelligent systems and knowledge engineering, we have put a step forward but still have a long way to go" at ISKE2010, Hangzhou, China.

He worked tirelessly for international series of conferences on ISKE until he passed away unexpectedly on July 31, 2011. His colleagues will continue his effort to make ISKE conference more successful, promote the study of intelligent systems and knowledge engineering in the future and make a greater contribution to humanity.

2.7 Guest Editor in Special Issues of Journals

1. Montero J, Ruan D (2010) Guest editors, special issue on Modeling Uncertainty, Information Sciences 180 (6), Elsevier

2. Montero J, Ruan D (2010) Guest editors, special issue on Soft Information Analysis, International Journal of General Systems 39 (3), Taylor & Francis

3. Lu J, Ruan D, Zhang G (2010) Guest editors, special issue on Intelligent Decision Support and Warning Systems, Knowledge-Based Systems 23 (1), Elsevier

4. Zeng X, Ruan D, Koehl L (2008) Guest editors, special issue on Intelligent Sensory Evaluation: Concepts, Implementations, and Applications, Mathematics and Computers in Simulation 77 (5-6), Elsevier

5. Lu J, Ruan D (2007) Guest editors, special issue on Intelligent Knowledge Engineering Systems, Knowledge-Based Systems 20 (5), Elsevier

6. Lu J, Ruan D, Zhang G, Zimmermann HJ (2007) Guest editors, special issue on E-commerce Intelligence, International Journal of Intelligent Systems 22 (5), John Wiley & Sons

7. Ruan D, Hines JW, Pázsit I (2005) Guest editors, special issue on Computational Intelligence in Nuclear Applications, Progress in Nuclear Energy 46 (3-4), Elsevier

8. Ruan D (2004) Guest editor, special issue on Information Fusion and Decision-Making under Uncertainties, International Journal of General Systems 33 (2-3), Taylor & Francis

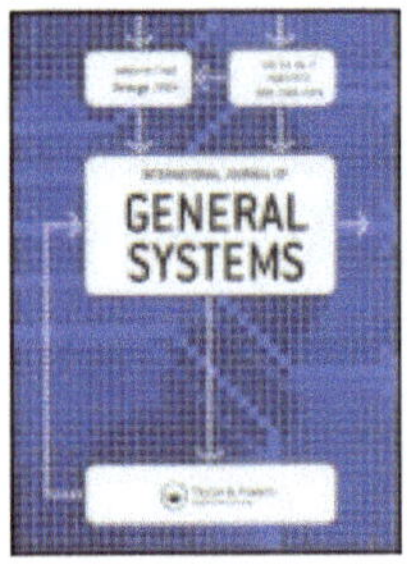

9. Ruan D (2004) Guest editor, special issue of Applications on Intelligent Real-time Systems for Nuclear Engineering, Real-Time Systems 27(1), Kluwer

10. Ruan D, Zhou C, GuptaMM (2003) Guest editors, special issue on Fuzzy Set Techniques for Intelligent Robotics, Fuzzy Sets and Systems 134 (1), Elsevier

11. Ruan D,Fantoni PF (2002) Guest editors, special issue on Intelligent Systems for Process Monitoring and Diagnosis, International Journal of Intelligent Systems 17 (8), John Wiley & Sons

12. Ruan D (2002) Guest editor, special issue on Intelligent Information Systems and Applications, Information Sciences 142 (1-4), Elsevier

13. Ruan D (2000) Guest editor, special issue on FLINS'98, International Journal of General Systems 29 (1/2), Gordon and Breach Science Publishers

14. Ruan D,Van der Wal AJ (1999) Guest editors, special issue on Robotics Applications at FLINS'98, Robotics and Autonomous Systems, 28 (1), Elsevier

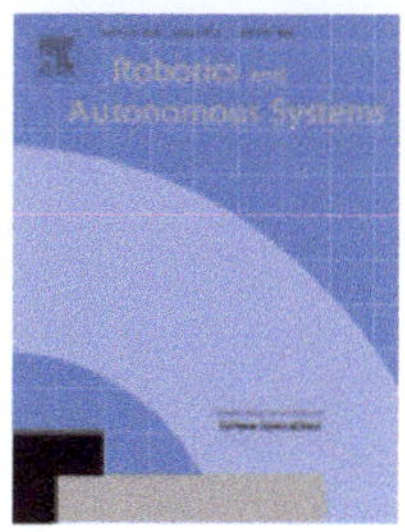

15. Ruan D (1998) Guest editor, special issue on FLINS'96, International Journal of Intelligent Systems, 13 (2/3), John Wiley & Sons

16. Ruan D (1997) Guest editor, special issue on Nuclear Safety, Korean Journal of Fuzzy Logic and Intelligent Systems 7 (1)

17. Ruan D (1995) Guest editor, special issue on Nuclear Engineering, Fuzzy Sets and Systems 74 (1), Elsevier

2.8 Keynotes, Tutorials and Panels in Conferences

2.8.1 Keynote Speaker in International Conferences

- May 12, 2012, invited keynote speaker at the 3rd International Conference on Quantitative Logic and Soft Computing (QL & SC2012), Xi'an, China
- June 11, 2011, keynote speaker on "Uncertain information analysis and intelligent decision support in perception based complex systems" at the 2011 IEEE International Conference on Computer Science and Automation Engineering (CSAE2011), Shanghai, China
- December 20, 2009, keynote speaker on "Perception-based decision analysis support in complex systems" at the 2009 International Workshop on Massive Data Mining and Knowledge Discovery (MDMKD2009), Chengdu, China
- July 19, 2008, keynote speaker on "The Role of Computational Intelligence in Complex Decision Systems" at the 1st International Workshop on Nonlinear Dynamics and Synchronization (INDS08), Klagenfurt, Austria
- October 16, 2007, keynote speaker on "Decision Support Systems in Society and Policy Support Applications" at the 2nd International Conference on Intelligent Systems and Knowledge Engineering (ISKE07), Chengdu, China
- September 26, 2007, keynote speaker on "Intelligent decision and policy making support systems" at the 1st International Conference on Risk Analysis and Crisis Response (RACA07), Shanghai, China
- September 14, 2007, keynote speaker on "Lessons learned from fuzzy logic applications in complex systems" at the International Conference on Fuzzy Logic and applications (EUSFLAT07), Ostrava, CZ
- May 23, 2006, keynote speaker on "Applied AI for Knowledge Engineering" at the University of the State Mexico Engineering faculty's 50th anniversary conference, Toluca, Mexico
- April 7, 2006, keynote speaker on "Applied Artificial Intelligence in Complex Systems," International Conference on Intelligent Systems and Knowledge Engineering (ISKE2006), Shanghai, China
- August 31, 2005, keynote speaker on "Artificial Intelligence in Nuclear Engineering: developments, lessons learned and future directions" at the International Nuclear Atlantic Conference (INAC2005), Santos, SP, Brazil

- July 22, 2005, invited featured speech on "Applied AI in Complex Systems: Examples and Lessons Learned" at the 8th Joint conference on Information Sciences (JCIS), Salt Lake City, Utah, USA
- September 16, 2004, keynote speaker on "Computational intelligence for control and decision making – from the viewpoint of practical applications" at the Spanish Conference of Technology and Fuzzy Logic, September 15-17, 2004, Jaen, Spain
- December 18, 2003, keynote speaker on "The role of computational intelligence in sensory evaluation" at the 1st Sino-French workshop on Sensory Evaluation and Intelligent Techniques, Shanghai, China
- September 27, 2003, keynote speaker on "Computational intelligence for applied research" at the 7th Joint conference on Information Sciences (JCIS), Cary, NC, USA

2.8.2 Invited Tutorials in International Conferences

- December 15, 2011, invited tutorial at ISKE2011, Shanghai, China
- November 14, 2010, invited tutorial on "Intelligent decision analysis support under various uncertain information in perception based complex systems" at ISKE2010, Hangzhou, China
- November 16, 2008, invited tutorial on "Computational intelligence systems and applications" at ISKE2008, Xiamen, China
- December 6, 2005, invited tutorial (6 hours) on "Tips for implementing applied AI in complex systems" at AI2005, Sydney, Australia
- September 9, 2003, invited tutorial (3 hours) on "Lessons learned from computational intelligence in complex systems" at the 3rd European fuzzy technology conference (EUSFLAT 2003), Zittau, Germany

2.8.3 Panel Organiser in International Conferences

- May 25, 2011, moderator of a panel on Soft Computing Challenges and Future Directions in Society and Industry, at the 2011 World Conference of Soft Computing, San Francisco State University, San Francisco, USA, May 23-26, 2011

  Background

  - Over the last two decades there has been a growing interest in the need for designing intelligent systems to address complex control and decision making problems. One of the most challenging issues for an intelligent system is to effectively handle real-world uncertainties that cannot be eliminated. Computational intelligent systems, based on Soft Computing techniques, including fuzzy logic, neural networks, genetic algorithms and others, have shown great potential to solve these demanding, real-world problems that exist in uncertain and unpredictable environments. To meet these research needs in Society and Industry, the panel will provide a comprehensive review of soft computing techniques and their ongoing implementation, focus on challenges from lessons learned and future directions, and discuss options for its further promotion in both education and research in general.

Moderator

Da Ruan, SCK•CEN www.sckcen.be (contact druan@sckcen.be)

Panelists

Piero P. Bonissone, GE Global Research (bonissone@crd.ge.com)
Oscar Castillo, Tijuana Institute of Technology (ocastillo@tectijuana.mx)
Ashok Deshpande, University of Pune, India (ashok_deshpande@hotmail.com)
Dimitar Filev, Ford Motor Company (dfilev@ford.com)
Jie Lu, University of Technology, Sydney (jielu@it.uts.edu.au)
Luis Magdalena, European Centre for Soft Computing, Spain (luis.magdalena@softcomputing.es)
Javier Montero, Complutense University of Madrid (monty@mat.ucm.es)
Witold Pedrycz, University of Alberta (pedrycz@ee.ualberta.ca)
Rita Ribeiro, Universidade Nova Lisboa (rita.a.ribeiro@gmail.com)
Lotfi A. Zadeh, University of California Berkeley (zadeh@eecs.berkeley.edu)

2.8.4 Invited Short Courses

- 2009 August 23-29, invited short course (5 hours) on "To compensate or not to compensate: fuzzy aggregations in complex decision systems" at the 13th Ph.D. Summer School on Advanced Course on Artificial Intelligence (ACAI2009), Belfast, University of Ulster, Northern Ireland, UK. Event sponsored by ECCAI and supported by the SGAI of BCS.
- 2009 April 9-10, invited short course (16 hours) on "Intelligent control" at the University of Ibague, Colombia for MS students
- 2005 February 11-12, invited short course on "New trends on intelligent systems and soft computing" at the University of Granada, Spain for PhD students
- 2003 December 20-24, invited short course (20 hours) on "Applied computational intelligence" at the Southwest Jiaotong University, Chengdu, China

- 2003 June 14-25, invited short course (10-day) on "Intelligent adaptive systems for nuclear applications: methods, design and implementations," The National Institute for Nuclear Research of Mexico, Toluca, Mexico
- 2002 December 15-21, invited short course (one week) on "Applied computational intelligence" at the Southwest Jiaotong University, Chengdu, China
- 2000 October-November, invited short course (8 hours) on "Soft computing" for PhD students at University of Antwerp, Belgium

2.8.5 Invited Guest Lectures

- November 15, 2008, invited lecture on "Soft computing techniques in control and decision making" at Xihua University, Chengdu, China
- May 15, 2008, invited lecture on "Fuzzy control" at Shanghai Maritime University, Shanghai, China
- April 5, 2006, invited lecture on "Computational Intelligence in Management Science" at Donghua University, Shanghai, China
- November 23, 2006, invited lecture on "How to write high quality journal papers" at University of Technology Sydney, Australia
- May 24, 2005, invited lecture on "Applied AI" at Donghua University, Shanghai, China
- July 14, 2004, invited lecture on "Applied computational intelligence" at Xihua University, Chengdu, China
- July 7, 2004, invited lecture on "Data mining applications" at Southwest Jiaotong University, Chengdu, China
- June 30, 2004, invited lecture on "Data mining and applications" at Donghua University, Shanghai, China
- June 24, 2004, invited lecture on "Data mining and decision making" at Fudan University, Shanghai, China
- December 26, 2003, invited lecture on "Lessons learned from computational intelligence in nuclear applications" at Institute of Nuclear Energy Technology, Tsinghua University, Beijing, China
- December 25, 2003, invited lecture on "Applied computational intelligence in complex systems" at School of Economics, Tsinghua University, Beijing, China
- December 25, 2003, invited lecture on "Applied computational intelligence and nuclear risk analysis" at Beijing Normal University, Beijing, China
- December 20-24, 2003, invited lecture on "Applied computational intelligence" at the Southwest Jiaotong University, Chengdu, China

- December 19, 2003, invited lecture on "Computational intelligence and its applications" at Tongji University, Shanghai, China
- March 2003, invited lecture on "Recent Developments and Future Research Directions of Applied Computational Intelligence" at the Beijing Normal University and China Academia, Institute of Software, Beijing, China
- December 2002, invited lecture on "Industrial Applications of Computational Intelligence" at Fudan University, Shanghai, China
- October 2002, invited lecture on "Applications of Soft Computing" at Ecole Nationale Supérieure des Arts et Industries Textiles, Roubaix, France
- May 2002, invited lecture on "Fusion of Hard and Soft Computing Applications for Nuclear Power Plants" at Ghent University, Belgium
- October 2001, invited lecture on "Fuzzy Logic and Soft Computing Applications" at Chalmers University of Technology, Göteborg, Sweden
- March 2001, invited lecture on "Fuzzy Techniques in Control & Decision-Making" for MS students, Ghent University, Belgium
- November 2000, invited lecture on "Safe Experiments of On-line Fuzzy Control for Nuclear Reactor Operation" at ANS 2000, Washington DC, USA
- May 2000, invited lecture on "Soft Computing: Applications in Power Plants" at the seminar on AI methods in Nuclear Energy, Prague, Czech
- April 2000, invited lecture on "Recent Development of Intelligent Systems and Soft Computing: Challenges and Future Directions" at the Southwest Jiaotong University, Chengdu, China
- April 2000, invited lecture on "Fuzzy Logic Control: Theory and Applications" at the Nuclear Power Institute of China, Chengdu, China
- December 1999, lecture on "The Potential of Intelligent Systems and Soft computing Applications in Industry" at SCK·CEN, Mol, Belgium
- September 1999, invited lecture on "Soft Computing at the Institute of Image Processing and Pattern Recognition" at Shanghai Jiao Tong University, China
- December 1998, invited lecture on "Advanced Fuzzy Control for Simulating a Nuclear Reactor Operation" for AI seminar at the Katholieke Universiteit Leuven, Leuven, Belgium
- September 1998, invited lecture on "The Potential of Fuzzy Logic and Intelligent Systems in Nuclear Science" for China-Flanders video conference to Shanghai, Gent, Belgium
- December 1997, invited lecture on "Intelligent Hybrid Systems and Industrial Applications" at TNO Physics and Electronics Lab, Den Haag, the Netherlands
- November 1997, invited lecture on "From Fuzzy Set Theory to Intelligent Hybrid Systems" for AI seminar in the Katholieke Universiteit Leuven, Leuven, Belgium

- October 1997, invited lecture on "Recent Developments in Intelligent Hybrid Systems" at International Colloquium on New Technologies based Learning and Employment Support, Leuven, Belgium
- April 1997, invited lecture on "Fuzzy Logic in Nuclear Engineering" at Belgium/Fuzzy'97 workshop, Mons, Belgium
- October 1996, invited lecture on "Fuzzy Logic Applications in Nuclear Research Areas" at the ESARDA-INMM'96 Workshop, Arona, Italy
- October 1996, invited lecture on "Recent Development of Intelligent Systems" at the Beijing Normal University, Beijing, China
- June 1996, invited lecture on "R&D on Fuzzy Logic Applications" at SCK·CEN at NAFIPS'96, Berkeley, California, USA
- February 1996, invited lecture on "Fuzzy Control Applications" at Centre for Systems Engineering and Applied Mechanics CESAME, Université Catholique de Louvain, Louvain-la-Neuve, Belgium
- November 1995, invited lecture on "Fuzzy Logic Applications for Decision-Making and Control" for AI seminar at the Katholieke Universiteit Leuven, Leuven, Belgium
- September 1995, invited lecture on "Fuzzy Logic Applications" at SCK·CEN, Mol, Belgium
- August 1995, invited lecture on "Fuzzy Information Knowledge Engineering" at ICIK'95 international conference, Dalian, China
- May 1995, invited lecture on "Applications of Fuzzy Systems and Neural Networks" at SCK·CEN, Mol, Belgium
- March 1995, invited lecture on "Theory and Applications of Fuzzy Logic" at the Nuclear Power Institute of China, Chengdu, China
- March 1995, invited lecture on "Fuzzy Logic and Intelligent Technologies in Nuclear Science" at the International Workshop on Fuzzy Engineering during the 4th FUZZ-IEEE/IFES'95, Yokohama, Japan
- November 1993, invited lecture on "Possible Fuzzy Applications for Safeguards" at IAEA, Vienna, Austria
- July 1993, invited lecture on "Fuzzy Approaches for Nuclear Safety" at the Institute of China Atomic Energy, Beijing, China
- July 1993, invited lecture on "Advanced Fuzzy Set Theory and Its Applications" at China Textile University, Shanghai, China
- May 1993, invited lecture on "Fuzzy Systems in Nuclear Applications" at SCK·CEN, Mol, Belgium
- April 1993, invited lecture on "Basic Principles of Fuzzy Set Theory to Engineers" at SCK·CEN, Mol, Belgium

- November 1992, invited lecture on "Fuzzy Decision-Making Theory" at Fuzzy Logic Conference, Holiday Inn, Antwerp, Belgium
- October 1992, invited lecture on "Fuzzy Logic and Decision Theory in Nuclear Science" at SCK·CEN, Mol, Belgium
- June 1991, invited lecture on "Fuzzy Set Theory and Practice" at SCK·CEN, Mol, Belgium

Part III
Da's Academic Activities

3.1 An Outstanding Scientist in Applied AI for Nuclear Research

Dr. Pierre D'hondt and Prof. Dr. Eric van Walle

SCK•CEN, Belgian Nuclear Research Centre, Mol, Belgium

After his PhD at Ghent University, Da was attired by the Nuclear Research Center of Belgium in 1991 to optimize a research program for oil reserves using nuclear techniques. His successful accomplishment led to an additional post-doc about 'Fuzzy Logic and Intelligent technologies in Nuclear Science'. His baby FLINS was born in 1994 and is since then being organized every 2 years somewhere in the world.

Prof. Dr. Da Ruan with the SCK•CEN management hosting a Chinese Delegation in 2006.

This immediately showed Da's organization-skills and on top of that, his exceptional quality to build a high-level scientific network by perseverance and charisma. He could bring leading world scientists together to discuss.

Since 1996, Da had a permanent position at the Centre and kept scientifically busy: he applied fuzzy logic, uncertainty analysis, information processing and other mathematical techniques to make decisions in a multitude of nuclear domains. I name a few of them:

- radiation protection: on how to preserve people and environment against radiation;

- the safeguarding of sensitive nuclear materials, which could possibly fall into terrorist hands;
- the design of a reactor control system for the BR1 research reactor in Mol and the TRIGA reactor in Mexico;
- the accomplishment of intelligent decision systems for the cooling of nuclear reactors. He developed this in Halden, Norway, where he and his family stayed for 18 months;
- cost-analysis of big and complicated projects;
- nuclear safety culture and social acceptance of nuclear energy: the domains on which he concentrated now.

Da published at large in highly recognized scientific journals; some years he was responsible for half of the scientific output of the nuclear research centre. He wrote several books, contributed chapters to more than one hundred volumes and was chief-editor or co-editor of some 50 scientific books or special editions of scientific journals. He was a highly esteemed invited and keynote speaker at international form and, in time, he belonged to some hundred international conference committees that made the program selection for top level scientific events. He was also referee for some 20 international journals and, last but not the least, guest professor starting from Ghent and Hasselt up to universities in China and Australia.

In 2006, Da was appointed as our Liaison Officer for China. From that time on, Da assured, on top of his other activities, our communication channels with China on the highest level: he had entries to the Embassies, facilitated difficult discussions, and mainly due to his devotion, we were able to have the Chinese Vice-Prime Minister Zeng Peiyan to visit the Centre late 2006. Another direct consequence of his Liaison task was the signature of multiple technical cooperation agreements between our Centre and renown Chinese institutes during the Royal Visit of the Belgian Crown Prince Philippe to China in June 2007. The last event Da co-organized was the Belgian Nuclear Day at the Shanghai World Expo in 2010, where important steps were taken to obtain Chinese support for the MYRRHA-system. Da brought life to these agreements, tried to push everyone to commitments, helped to enlarge some cooperation on the European Commission level and he followed them up until one week before passing away.

Needless to say that Da acted as an international magnet for tens of stage airs, PhD and postdoc students. Da was severe for himself, a perfectionist on most aspects, but was also very demanding towards his students. They could count on his gentle kind at all times and for all matters, but they had to comply with many of his scientific desiderata: he educated them well and according to his principles.

The diversity of research topics that Da treated within SCK•CEN, made him active in different research teams and sometimes he was looked upon as a 'free electron' that made its contributions to the Centre as a whole. This was not easy, not for his hierarchy, not for Da himself, although he always persevered positively and effectively realized his assignment or idea.

This multitude of professional contacts of Da inside and outside of SCK•CEN resulted in a huge circle of acquaintances; acquaintances who often became friends due to his warm, social attitude. Da knew very well what it was like to arrive as a stranger in a new country and to be taken care off. Both in Belgium and

China, Da was a humble cultural ambassador for his beloved country and the Chinese community.

With the passing away of Da, the nuclear Centre has lost one of its most brilliant researchers and scholars. He had so much to offer and still had so many plans. Although one always says that everyone can be replaced, Da cannot. We will have to reorient his research and will try to do so by keeping in mind his philosophy of openness, hard work and perseverance.

The unexpected happened so sudden and disguised that no one was prepared to counteract. Dear Da, your colleagues at SCK•CEN, the institution that you carried in your heart, will commemorate you as our pioneer, but also as a worldwide recognized expert in soft computational techniques. Above all we want to thank you for your kind and warm personality that we will miss every day.

3.2 Da and the Chinese Community of Mol

Dr. Hong Chen

VITO, Mol, Belgium

Mol is a small town with about 35,000 inhabitants, located in the Belgian province of Antwerp. This is the city where Da Ruan had worked and lived with his family and friends for the last 20 years.

In 1991, Da finished his PhD at the University of Gent and started to work at the Belgian Nuclear Research Center (SCK-CEN) in Mol.

As Da's one of the oldest Chinese friends, I had a similar study and work experience (Chinese student studied abroad in the 80s). I started to work at SCK-CEN in 1989.

20 years ago, we were so young, still singles, carefree, energetic, optimistic, enthusiastic ... together with some other young scientists from Belgium and other countries, we became friends and colleagues. In our spare times, we used to live in the same dormitory building provided by our employer, SCK-CEN. We organized all kinds of activities: parties, cooking lessons, going to cinema, restaurants, bars, making trips, playing cards, sport, visiting local friends and families...... That was really our golden time with unforgettable experiences.

Da playing cards during the Chinese party in 2003

Karaoke during the Chinese party in 2006

Soon later, Da and I and a few other Chinese friends started our own Chinese community because more and more Chinese scholars and families started to work and live in Mol. From time to time, there were Chinese people coming and leaving, but our Chinese community was always kept alive. Da was extremely supportive and loved all kinds of new ideas to improve our Chinese activities.

A few years later we both started our own family and got children, therefore the activities were limited only once or twice per year for celebrating the Chinese new year in winter and the Moon festival in autumn. All the Chinese families around Mol loved those gathering parties, because the parties were always organized in the pleasant ways: nice place, happy atmosphere, plenty of delicious food, beautiful decorations, interesting games and sometimes having also Karaoke. The idea of each person or family prepared one or more dishes were highly appreciated. The dishes Da prepared were the "white-cut chicken", a typical dish with Shanghai style and "CaoFu" a typical NingBo dish. They were so delicious therefore they were awarded as one of the traditional dishes kept for our Chinese parties. The best part for us was playing cards with Da at the end stage of the party. The gathering parties were so successful and all the participants enjoyed the party and looked forward to having the next ones. Da was always the main organizer and responsible for renting the place, so that our Chinese community considered him as a symbol of Chinese parties. This reminds me a poem: "Mol became popular through you and now is a place that we will always cherish (Prof. Dr. Jie Lu, in the memory of Da Ruan, on 6 Aug 2011 Mol Belgium)". For our Chinese community of Mol, because of Da, we feel so proud that we had been part of Mol and part of Da's life in Mol.

We, our Chinese community cannot forget that we had spent great time with Da for the last time in the Chinese party held in February 2011. Da and I had proposed to have the next party in autumn to celebrate the Chinese Moon day. There was no party unfortunately in the autumn 2011 because Da left us, because we were so sad. We missed you, Da!

3.3 An Excellent Guest Professor at Ghent University

Drs. Chris Cornelis[1,2], Martine De Cock[1], Glad Deschrijver[1], Mike Nachtegael[1], and Steven Schockaert[3]

[1] Ghent University, Ghent, Belgium
[2] University of Granada, Granada, Spain
[3] Cardiff University, Cardiff, UK

Prof. Dr. Da Ruan was officially appointed as a guest professor at the Department of Applied Mathematics and Computer Science, Ghent University, since October 1, 2005. However, as many former students will be able to confirm, Da was already deeply involved in research and teaching at the university long before receiving the official title of guest professor. In this chapter, five of those former students reminisce about Da's great influence in their professional lives as well as that of other students and researchers at the department.

Guest Lectures

Da loved to teach and share his knowledge with students. He was a very welcome guest lecturer in graduate courses on computational intelligence, where he would teach with great credibility about the application of soft computing techniques in practice. His lectures in courses such as "Guest Lectures on Current Aspects in Computer Science", "Capita Selecta from Fuzzy Set Theory" and "Computational Intelligence" stood out and left deep impressions on students in Computer Science and Mathematics at Ghent University.

Steven Schockaert recalls: "I particularly remember a guest lecture Da gave to the students of the Computational Intelligence course, about the use of fuzzy control in nuclear science. Rather than giving a standard lecture on the advantages of fuzzy controllers, he presented a very balanced story, addressing both virtues and vices of fuzzy controllers, and fuzzy methods in general. Clearly he was in a unique position to comment on this, being both involved in fundamental research as well as in industrial applications. His continued efforts to find new ways in which the theory of fuzzy sets could be applied in nuclear science will surely have a lasting impact on the field."

Martine De Cock recalls: "In one of the lectures that I attended, Da stressed how difficult it is to put a new method based on fuzzy systems into practice in a nuclear application, because of strict regulations w.r.t. safety guarantees of new methods. He told the class that for this reason, there is a tendency to hold on to the old methods, even when they are more costly. This was an eye-opener for me because, until that point, I had always assumed that a new method with a clear potential to save money and resources would immediately be embraced… The clearest evidence of Da's love and devotion to teaching was perhaps the fact that he never seemed to hesitate to take time out of his very busy schedule at SCK-CEN – the Belgian Nuclear Research Centre – and travel the long distance from Mol to Ghent each time to teach at Ghent University. We are really going to miss him as a valued guest lecturer and a source of inspiration for new generations of students."

Master Thesis Students

Throughout the years, Da's involvement at Ghent University was of great influence on the master theses of many graduate students in Mathematics as well as Computer Science, and often even the start of their scientific career.

Mike Nachtegael (master thesis in Mathematics, 1997-1998): "Although I was not a master thesis student of Da, his work has had a great and crucial impact on the start of my scientific career! During Prof. Etienne Kerre's course on Fuzzy Set Theory, we used the book 'Introduction to the Basic Principles of Fuzzy Set Theory and some of its Applications'. I became very much intrigued by this new world that opened up to me. In that book there was a chapter dedicated to just one type of logical operators: the implication! That chapter was written by our good friend Da. I was so fascinated by it that I proposed to do my master thesis on that subject. Prof. Kerre immediately agreed, and what followed was a thorough journey in the world of implications. It was my chance to show my scientific capabilities, which finally resulted in Prof. Kerre's initiative to let me join his research group as a PhD student. I do not know how things would have worked out – and if I would still be active at Ghent University now – if it wasn't for that great chapter of Da in that book…"

Chris Cornelis (master thesis in Computer Science, 1999-2000): "In 2000, while I was working on my master thesis, Da offered me the possibility to write an overview chapter on the generalized modus ponens for a book he was editing together with Prof. Etienne Kerre. I vividly remember the sense of pride I felt that a paper I had contributed to, would actually be printed and published in a real book."

Through his appointment as guest professor, Da actively stimulated master theses and internships in cooperation with other SCK-CEN members. For instance, Sofie Van Landeghem (master thesis in Computer Science, 2006-2007) was co-supervised by Dr. Rafi Benotmane from the Laboratory of Radiobiology, and Bert Bonte (master thesis in Computer Science, 2006-2007) was co-supervised by Paul

Borgermans from the ICT department of SCK-CEN. These cooperations would not have been possible without Da's altruistic efforts to bring interested parties together.

In addition, Da also offered inspiring thesis topics for master students himself. A very nice example is Da's supervision of Patricia Victor (master thesis in Computer Science, 2004-2005). Initially, Patricia was very hesitant about whether she wanted to engage in scientific research after her graduation. However, during 2004-2005 she regularly traveled to Mol to work with Da on her master thesis, and gradually got more and more interested. She successfully applied for a PhD-scholarship with the Flemish Agency for Innovation by Science and Technology. Her PhD thesis on 'Trust Networks for Recommender Systems' was published in 2011 in the Atlantis Press Computation Intelligence Series, edited by Da.

Glad Deschrijver: "Da was until the last day of his life much involved in research and teaching and not a hair on his head (which was so young that it had not lost any yet) was thinking of slowing down. This is exemplified among other things by the fact that my current master thesis student, Lotte Muylaert (master thesis in Mathematics, 2011-2012), originally planned to do a master thesis with Da on applications of aggregation operators. Although she only visited him once and he had not yet explained the subject in depth, she knew that she wanted to do her thesis with Da. His sudden death changed these plans."

PhD Students

Through his frequent visits at Ghent University, and his enthusiastic and outgoing nature, Da inspired many students in the course of their PhD work. This is particularly visible in the PhD of Yun Shi (PhD in Mathematics, 2010) for whom he acted as a devoted co-supervisor and mentor; her thesis on fuzzy implications completed, in a sense, the work he had initiated years before in the course of his own PhD. Throughout the years, Da also served on many PhD exam committees; his questions at the defenses showed how he read the works with deep understanding and interest. When reflecting on Da's influence in their scientific career, former PhD students also often especially recall the FLINS conferences. Through the organization of these conferences in Belgium, Da provided a wealth of important close-to-home learning experiences w.r.t. presentation of scientific work, organization, and establishing a professional social network.

Martine De Cock (PhD in Computer Science, 2002): "I have always admired Da for his diplomatic interpersonal problem solving ability, and his great skills as a marketer. He taught me that, even when the goal of our research is to develop automated systems that can think like – and thus replace – humans, presenting our work as such might be met with a hostile attitude from humans who dislike the idea of 'being replaced'. An idea that Da was particularly fond of in this respect were 'decision support systems': systems that aid humans in their decision processes, rather than replacing them."

Da and Mike at Mike's PhD defence reception in 2002

Mike Nachtegael (PhD in Mathematics, 2002): "My very first publication of a scientific article as a PhD student was at the FLINS 1998 conference, organized by Da Ruan! As mentioned earlier, I dedicated my master thesis to fuzzy implications, inspired by the chapter on implications that Da wrote for Etienne Kerre's course book. After finishing that master thesis I continued with some more research on implications on finite chains. In particular I calculated the number of different types of implications on such chains. The results of this work were written down in my first paper. The presentation at FLINS 1998 also was my first 'live' presentation experience. These are really memories that had a great influence on my scientific life, and for which I will always be grateful to Da."

Martine De Cock (PhD in Computer Science, 2002): "At EUSFLAT2001 in Leicester (UK), I was in Da's company during one of the lunch breaks. We went to a pizza restaurant, hoping for a quick meal. The service was however very slow. Da started analyzing the organization in the restaurant, and estimating how much more money could be earned with a more efficient approach. Later during that same lunch, when the topic of FLINS came up, Da asked if I would be interested to help organize FLINS2002 in Ghent. As a PhD student with very little experience, I had no idea what I was getting myself into when I agreed. But, since the request came from a man who had just critiqued and suggested an improvement of the organization of a restaurant, I did expect that it would be a lot of work and at the same time a great opportunity to learn from someone with a lot of experience. This is exactly what FLINS2002, and later FLINS2004 and FLINS2006 turned out to be."

Steven Schockaert (PhD in Computer Science, 2008): "Although Da was not directly involved with my PhD, he has influenced it in many ways, starting from my initial choice to do a master's thesis on an application of fuzzy set theory. As a third year student, I had a piece of coursework on inference in fuzzy logics (in the narrow sense), for which I was given a paper by Da on fuzzy implications as reference material. This sparked my initial interest for the idea of fuzzy sets and multi-valued logics. The year after I worked on a master's thesis which involved the use of Mamdani inference in a clustering algorithm, and I wrote my first paper, which I presented at the FLINS 2004 conference. It is at this conference where I met Da for the first time, in the bar of the hotel after the first conference day, where he was talking enthusiastically about the fact that the next Summer Olympics were going to be held in Beijing (the 2004 Summer Olympics in Greece had just finished a few days before). I was immediately charmed by this positive energy and optimistic attitude. I particularly remember the last day of the conference, which saw very nice weather that had led many of the conference attendants to the beach and to the city of Bruges. Da responded to this by promising to buy dinner for everybody who had stayed until the end of the last session at the next FLINS edition. Finally, at the end of my PhD, it was Da who personally convinced me of submitting my thesis for possible publication to World Scientific. As the series editor of the 'Intelligent Information Systems' series where the thesis was eventually published, he has been of enormous help to organize the publication of the book, and to speed-up the review process. I spoke to Da for the last time at the 'Christmas lunch' at Ghent University in January 2011, where we talked for a long time about my plans to apply for a lectureship position abroad, and where he gave me a lot of good advice. Looking back, it is only now that I fully realize that Da has played such an important role in all the crucial steps of my academic career."

Visits of International Researchers

Da was extremely well connected in the scientific community. Many researchers referred to him not only as a colleague, but also as a friend. Da regularly invited international contacts to Belgium, and to Ghent University, thereby stimulating international cooperation. Among the most recent visits, still fresh in mind, were those by Prof. dr. Jian-Bo Yang from Manchester University (UK), dr. Lusine Mkrtchyan from IMT Lucca and University of Pisa (Italy), Prof. dr. Jie Lu from the University of Technology, Sydney (Australia) and Prof. dr. Vassilis Kaburlasos from the Technological Educational Institution of Kavala (Greece). In addition, Da provided tremendous support in the China-Flanders cooperation projects in which the Fuzziness and Uncertainty Modelling Research Group at Ghent University actively participated for many years.

Chris Cornelis: "When Da learned about my interest in recommender systems, he introduced me to Prof. Jie Lu at FLINS 2004, which lead to a fruitful scientific visit to her group at the University of Technology, Sydney."

Mike Nachtegael: "I can still picture the last talk that Da organized! It was a very nice presentation on recommendation systems by his good friend Prof. Jie Lu, who came all the way from Australia. After the talk, and after the typical 'parking problem', we went to a very nice Chinese restaurant. It was a great lunch, with a lot of conversation, memories and laughs. This was typical for Da, always the light at the table. It was the last time most of us saw him..."

Glad Deschrijver: "Da's death left many things unfinished. After the presentation of Prof. Jie Lu, he invited another speaker, Prof. Vassilis Kaburlasos, to Ghent to speak about a unified modeling and knowledge representation based on lattice theory. Unfortunately, this talk had to take place without Da."

At the time of writing of this text, we are preparing for another visit of Prof. Jie Lu to Ghent University. This visit has been made possible thanks to Da, who successfully applied for the financial support. It will be with great sadness in our hearts to see this last visit organized by Da take place without his warm presence, and, at the same time, we will smile, filled with many fond memories of a great man and researcher who so often went out of his way to help others on their path. We are grateful for everything you did for us and meant for us, Da. We will never forget you.

Da, Etienne and Jie surrounded by PhD students of Ghent University in Jan. 2005

3.4 An Excellent Guest Professor in Hasselt University

Prof. Dr. Geert Wets, Prof. Dr. Elke Hermans, Prof. Dr. Davy Janssens, Prof. Dr. Koen Vanhoof, Edith Donders, Qiong Bao, Yongjun Shen, and Dr. Banghua Yang

Hasselt University, Hasselt, Belgium

Geert's Memories

Prof. Da Ruan died completely unexpected on July 31 2011. Since September 2005, Da Ruan was a guest professor at IMOB. As a guest professor he was involved in several teaching and research activities. As part of his teaching activities he lectured on Knowledge-Based Systems and Computational Intelligence. As part of his research activities he worked closely with several IMOB researchers where his outstanding expertise regarding fuzzy sets and multi criteria analysis proved valuable w.r.t. their doctoral research.

Beijing Institute of Technology, June 2011

Furthermore, Da Ruan worked very hard to set up collaborations with several Chinese universities (e.g. Beijing Institute of Technology, Tsinghua University, etc.) and companies. This resulted in several research projects on the pre- and postdoctoral level. Finally, Da Ruan was the driving force behind international conferences such as FLINS and ISKE (of which the 2009 edition was organized at Hasselt University). With Da Ruan IMOB loses an excellent researcher and valuable colleague. Underneath you will find testimonials of several researchers at IMOB who worked closely together with him.

Elke's Memories

When working on my PhD at Hasselt University, I came in contact with Da. We had several productive meetings concerning missing data, expert weighing and aggregation operators, resulting in a number of papers. Da was always prepared to discuss research challenges in detail and came with good ideas. It was an honour to have him at my doctoral defence at which he asked me two and a half questions. In addition to the professional relationship, I consider Da as one of my best colleagues, always informing about the well-being of me and my family. Although we had a number of research plans for the future, I hope that by always remembering his enthusiasm, optimism and perseverance, some of our ideas will become reality in the future.

Da in toga at Hasselt University, April 2009

Davy's Memories

When I think of Da, many things can be said but there are two important characteristics of his personality which will always come into my mind. The first thing was his passion for his work. You could feel it in everything he did, in everything he said. I have met a lot of people who are working hard, very hard often, but I did not meet a lot of people who are as passionate as Da was. And there is an enormous difference between someone who is working hard and someone who is passionate: it is like adding a new dimension. He told me several times to define high standards for Ph.D. students. And a high standard for Da was not only translated in the traditional "number of publications", it was also hunting for projects, for new ideas, for innovation and for real scientific challenges. He had several contacts with Chinese researchers for projects, not for his own sake, or for his own career but for the benefit for science and overall knowledge. Da was a passionate person.

The second characteristic was his genuine interest in people. When Da asked you whether everything was fine, it was not just a common question for him, but he showed real personal interest. Numerous examples can be given which illustrate the often very practical small things he did to help others. Also people he did not know before, or did not work together with before. Da was a social person.

I will always remember Da of his great personality and I am sure he can be an inspiring source for many of us in the years to come.

Koen's Memories

To honour Da is not so difficult. He was lovely, friendly, social man. I can't remember when we first met. This is not surprising. Da was not a so called 'table jumper'. Probably we met during the bilateral research projects Flanders-China in the 90's. From the Flemish part the universities of Gent and Hasselt were involved and at that time Da was working part-time in Gent under the supervision of Etienne Kerre. We did common research and published together. I also started participating in the FLINS conferences. In these conferences you could see how social en beloved Da is in the research community. Da's friends and network are also very open. We appreciated the quality of his work and Da obtained a part-time position in our group. This was a new step in our relation. We were supervising PH.D students together, travelled together and organized the ISKE conference in Hasselt. Working with Da was a pleasure because it was high quality work done in a friendly, smoothly manner. So I repeat to honour Da is not so difficult, to accept his sudden dead is. Da thanks for everything, you will stay in my heart and I will special remember you when during free time I bicycle along the canal passing the SCK building.

Edith's Memories

As head of the IMOB secretariat I had the pleasure of meeting Da regularly. What the members of the secretariat will cherish the most is his thoughtfulness and humour. Da was IMOB's spokesperson in setting up collaborations with several Chinese universities and companies, which implied that he had to travel quite often. Da always turned to us for practical assistance. And he always made sure that he showed us how much he appreciated that help. He did this by simply thanking us at every occasion he had, but also by always buying each member of the secretariat a small present every time he went abroad. He made sure that he delivered those presents in person, always accompanied with a smile and an interesting travel story. Da was an example, not just because of his exquisite research work, but also because of his exquisite personality: always thoughtful, always respectful, always generous. We will miss his stories. We will miss his smiles. We will miss him and he will never be forgotten.

Qiong's Memories

In my opinion, Prof. Da Ruan is a respectable teacher and friend. I consider myself fortunate to work under his direction and benefit from his broad international experience. It's him who guides me into multi-criteria decision making field, and encourages me to proceed into my current research domain to further investigate the valuable points of transportation behaviour. I would like to give my deepest gratitude to him, for his invaluable support, gentle encouragement and constructive advice during the past years since I came to Belgium. As a friend, he is also memorable not only because he likes to share joy and happiness but also for that he is always ready to extend a helping hand. He always uses his wisdom and experience to help everyone handle practical problems and find out best solutions. Finally we all benefit from his kindness and learn from him about how to overcome difficulties and also how to get along with life.

Yongjun's Memories

Da was a guest professor of Hasselt University in Belgium, and part of his work is to supervise PhD students like me. In the past three years, Da entered my life as a respectable supervisor, an admirable senior, and also a close friend. All experiences with him were always impressive and memorable. As one of my co-promoters, we had meetings together every time when he came to the office, and he always brought me fresh ideas and distinct opportunities. I learned a lot from him, not only his rigorous scholarship, but also his enthusiasm, responsibility, and dedication. Based on his supervision, we had published nearly twenty papers together in different academic journals and conference proceedings during the past three years, and we also had a lot of plans for the future. As a senior of studying abroad, he was regarded as an outstanding example for us to follow. Despite his position, he had an unassuming personality. He always kept his words and did his best to meet his commitments. He generously shared his knowledge and experience without any reservation, and he was always ready to provide selfless help. As a good friend, I was also impressed by his kindness and openness. He invited me to his home tasting his special dishes. He joined our Chinese New Year party and made dumplings together with our Chinese students. We took part in group activities organized by IMOB and had dinners together at restaurants.

At the last academic year of my PhD study, he left us, so suddenly, without any portent, which means that he will be absent from my final defence, what he never did in my every three monthly PhD meeting before. But I know he will be there, in my heart, forever.

Da in the IMOB dinner in Hasselt, September 2008

Banghua's Memories

I successfully applied and obtained a post-doctoral position at IMOB of Hasselt University in Belgium under the recommendation of Prof. Ruan Da, which can also urge and strengthen the cooperation between Hasselt University and Shanghai University. During the application procedure of this post-doctoral position and also my VISA, Prof. Ruan Da did a lot of work and answered many of my questions. I knew he was very busy and I often received his emails at deep night. However, he thought everything carefully, and everything went smoothly because of his care and help. He was a passionate person and was very helpful to others. At the first day I came to Belgium, he worried that I did not know where to buy food. So, he brought me enough food and arranged some students to help me. It was not easy for him because he lived in another city and he always had a lot of work to do.

When I started my research in Hasselt University, Prof. Ruan Da was my associate supervisor. He discussed with me and other research members for several times when he came to University. In addition, we often communicated by emails. He always gave me many good ideas and guided me to a right direction. I learned much from him. He also asked me to join many research groups to discuss scientific problems, which could not only improve my English but also help me to have more friends. When he was at University, we often had lunch together. It would take us about fifteen minutes from the office to the restaurant and we would like to walk to cover this route. During the walking and eating, he told us many interesting things about his life in Europe and often encouraged us to overcome difficulties to do more meaningful things. I received Prof. Ruan's email two days before his passing away. He reviewed my manuscript carefully and gave many valuable suggestions in the last email. Before the last email, he had reviewed the manuscript for several times and gave many advices every time. His passing away was a great loss of my post-doctoral research and I lost an excellent supervisor and a good friend.

Both in my research and in my life, Prof. Ruan Da gave me great concern, support and help. To the memory of this respected promoter and distinguished professor, I would like to say:

没有你就没有我的今天
你对每个朋友和每个学生的友善
你的博学多才
给我留下了深刻的印象
你将永远活在我的心中
永远是我最好的导师
一位非常优秀的中国人
一位伟大的科学家
留给我永远的回忆

English version of Chinese poem:

I could not achieve so much in my research without you
I was impressed by
Your kindness to every friend and each student
Your broadness in learning and knowledge
You will always be alive in my heart as
My best promoter
An excellent Chinese
A great scientist
Memories of you will stay with me forever

3.5 In the Days since Da Passed Away: Our Best Adjunct Professor and Best Friend

Prof. Dr. Jie Lu, Prof. Dr. Guangquan Zhang, Dr. Jun Ma, Dr. Fengjie Wu, Dr. Chao Wang, Dr. Jack Zhang, and Julwan H. Purba

University of Technology, Sydney, Australia

Prof. Da Ruan was appointed as an Adjunct Professor at the Faculty of Engineering and Information Technology, University of Technology, Sydney (UTS), in 2006 and his appointment was renewed in 2010. In recent years, he took an active role in the research of the Decision Systems and e-Service Intelligence (DeSI) Lab, in the Centre for QCIS at UTS, including grant application, joint publication and PhD supervision. He was a partner investigator for two Australian Research Council (ARC) grants with us, and an external supervisor for two PhD students in the DeSI lab. He also collaborated on 20 joint publications with us. He is fondly remembered for a public seminar in 2006 entitled, "How to Write High Quality Journal Papers". He organised the agreement between UTS and SCK which directly helped our successful cooperation.

In early August 2011, the DeSI Lab established a web page in memory of Da Ruan: http://decide.it.uts.edu.au/DaRuan/Da_Ruan.html, which contains many friends' messages in tribute to Da's scientific work and his excellent character, and a collection of his photos. In this chapter, seven DeSI lab members reminisce about Da's great influence in their professional lives and his friendships with them.

Da, Guangquan, Jie, Chao, Jack, Fengjie and other friends in B 10, UTS in Nov 2006

Jie's Memories

1 Aug 2011

Monday, 1 Aug. 2011, 8am (Sydney time), checking emails as usual before going to the office, I turned on my laptop. An email sent by Prof. Kerre with the heading "sad news" was in the inbox. I opened it with hesitation and fear. "Dear friends and colleagues, I just heard the terrible news that my good friend Prof Dr Da Ruan suddenly passed away this early morning! He only became 50 years! Life can be very hard!" I could not believe my eyes. I read it three times. I cried out "Guangquan, Ruan Da left us!" "What?" Guangquan could not believe his ears. We then read this email several times together. It was sent by Prof. Kerre; the email address was correct.

We still could not believe it, because we had been in contact by email just a few days before and we knew Da was very healthy. We decided to call Prof. Kerre. My body, especially my hands, shook so badly that I could not even type the telephone numbers. We called twice but could not reach him. We decided to call Da's mobile phone. How desperately we hoped that this terrible news was not true! Da's wife, Greet, took my call. It was 1am, 1 Aug. in Belgium. We didn't believe it and we didn't want to believe it, but it was true—Our best friend Da Ruan left his family and his friends at 1am on 31 July 2011.

Our hearts were broken, we were deeply saddened, and we had no words, only tears streamed down our faces. Only 50 years old, he had no time to think, no time to prepare, no time to plan, and no time to say good-bye; he suddenly left us for his heavenly abode, and thus we belong to two worlds. I followed Greet's request to send this sad news to Da's friends in this world and to our community.

We decided to go to Belgium to attend the ceremony in memory of Da's life. Guangquan and I prepared two gifts for our best friend Da. One is a Chinese memorial plaque, suggested by Guoqing, and another is a commemoration book (photo album). Within two days, Guangquan collected signatures from 18 of Da's Chinese friends and put them on this plaque with the following Chinese poem to highlight Da's kindness, warm-heartedness, sunshine characteristics and generosity.

阮达兄弟走好！
你无比的才学，
无暇的人品，
无双的阳光性格，
留给我▯ 无尽的思念。

Within two days, we edited and printed a commemoration book which includes 14 selected pictures of Da and his short CV, to give to Da's wife, two sons, parents, sister, brother and some friends. Jun helped me with the format of this book. Emotions were constantly and strongly stirred throughout its preparation, so that we sometimes had to stop our work.

I departed on Thursday, 4 Aug. for Belgium with ten copies of this commemoration book and the Chinese memorial plaque. I had visited Belgium, including Mol and Ghent, several times and in almost all of those visits we had meetings with Da. In those previous trips, I relaxed or did some research preparation for our papers, book chapters or project proposals, during the long 24-hour trip. This time, I was going to say good-bye to Da—my heart was very heavy, my whole brain thought about Da, I could not stop recalling his stories ... could not sleep for a minute during the flight. It was very difficult to describe my feelings in those few days: huge shock, deep sadness …

I didn't think I would be able to see Da's body since the farewell to Da's remains was held at 4pm on Thursday 4 Aug. When I arrived at Brussels airport on Friday at 11:40am, Da's family friend, Ms Shao, picked me up and said, "Please help me to check the GPS. If we drive smoothly, you can see Da before his final departure at 1pm", Thank God, we arrived in Mol at 12:45pm. He was in eternal sleep. He still wore his glasses, grey shirt and yellow suit. I sobbed uncontrollably. "Da, how could you leave us?" "Why did you leave us?" "Were you too tired?" Providence is always hard to ask for—an incurable pain for our hearts.

Greet invited me to stay in their house. Da's sister Fei was there as well. Da's home garden was very green; it was summer. A traditional red Chinese lantern was strung in the trees around the lawn. A magnolia tree, the Shanghai City Flower, was just in blossom—for the first time, I was told. Da's laptop was still there. The bookshelves were filled with his published books and edited proceedings. Da's voice and expression were still fresh with me. I put the Chinese memorial plaque there and I burned a stick of incense in front of Da's photos: "Da, thank you for all your help, support and advice in the past years to me, my family and our friends, thank you for your long time friendship and distinguished leadership; thank you for everything you did for us. We are lucky to have been your friends. We are lucky to have worked with you. Please take care of yourself. We will miss you forever."

Burned a stick of incense in front of Da's photo

Luis was with me for the ceremony of Da's life in the SCK conference hall. Da's picture taken in Sydney was shown on the big screen in the hall—he was full of smiles and energy, which made me recall the wonderful time of his visit to Sydney in 2005 and 2006, and when I thought of his promise to visit us with his family at the end of 2011, my tears fell. Luis told me, "Be strong, Jie."

When I gave my speech in the ceremony, which was a Chinese poem I wrote with Guangquan and which was proofread by my parents, and its English version, I felt Da was with us. He was in the conference hall to listen. He saw how so many of his sweet friends and colleagues thought well about him and deeply missed him. He was full of smiles.

Meetings with Da

The first time I met Da was at FLINS2004, although Guangquan had talked to me about Da many times. At about 6pm on 1 Sept, 2004 I entered the dining room of the conference motel in Blankenberge, Belgium. I saw Prof. Hans Zimmermann was there having dinner by himself, and opposite him there was a plate full of food, but nobody was sitting there. When we were close to finishing our meal, a person came in, saying "Hi!" to almost everyone in the room with his sunny and pleasant smile; then warmly asked people to move to closer to Prof. Zimmermann's table, and finally sat down in his place. He was Da, the chair of the FLINS conference series.

Madan, Guoqing, Da, Javier, Jie and others in FLINS 2004

He had no time to eat because he had received an urgent call about a conference attendee who had lost his bag, containing his passport and wallet, in Brussels. He took care of this attendee just like he would look after a family member. Indeed, within a very short time, I proved that FLINS is just like an academic family and each member is Da's family member. At the conference, I saw how he was respected and loved by the FLINS family members and how he obtained such high academic achievements, and particularly, how he was always kind and helpful. I fully enjoyed his organisational skills and especially his true friendship.

I shall never forget our last meeting at Ghent University on 6 June 2011, eight weeks before Da passed away. I was invited by Prof. Kerre and Da to visit Ghent University as a visiting professor. I gave a seminar on Recommender Systems for e-Business Intelligence at 10:30am on 6 June. Da drove to Ghent from Mol to attend this seminar. After my seminar, we discussed our joint ARC grants and a joint special issue. In particular, Da told me about the organisation of the SSRAOC workshop and the progress of his journal, IJCIS. He told me his detailed plans and arrangements for ISKE2011 and FLINS2012 as well. He chose a Chinese restaurant for lunch with Prof. Kerre and Mike. During this lunch, he told us his plan for his next two trips to China and Spain, and finally asked for a pair of chopsticks from the restaurant for me to use in my hotel. Who could believe that it would become the last meeting with Da?

Last meeting with Da at Ghent on 6 June 2011

My visit in SCK, Mol

Da, "Mol became popular through you and now is a place that we will always cherish." Many friends in our community fully agree with the sentence in my poem. Mol is a small town, having a population of only 30,000, but over the last twenty years, Da attracted more than 100 visitors to SCK for research cooperation with him and his team, and I am one of those visitors.

In 2006, I received an Australian Academy of Science Award for my project 'Dynamic decision making with uncertain information processing in emergency management' under The Scientific Visits to Europe Program. I visited Mol in Jan 2007. Da warmly introduced me to his colleagues and showed me around his building and his office. I was very surprised by a mechanistic model which showed how Da successfully applied fuzzy logic techniques to a nuclear reactor. He had a patent on it. I was also surprised by the mass of books and papers in his office. I quickly realised how much work he took on and what a heavy workload he had. He was a person who would always aim for top-notch performance and never stopped working. Da, for you, one day has more than 24 hours.

In Da's office, the most obvious thing is the big framed photograph on the wall in which Da's two sons Alexander and Nicolas appear so happy. Da told me many of their stories and I can see that he devoted all his love to his sons. He also told me in 2011 how Nicolas studies C++ programming language on the Internet even though he is only 10 years old.

When I asked if his workload was too high, Da told me that he loved what he did, liked challenges and that all the work he did should be truly excellent. He also told me that he often had a beginner's mind for learning new things. He had a very effective working style. During my visits, we completed a joint journal paper and finalised our joint book "Multi-objective group decision making: methods, software and applications with fuzzy set technology" (Imperial College Press) which has been highly cited since publication. More importantly, I learnt a great deal from Da during my visit, much more than I expected and also much more than Da was aware.

It's a great pity that I didn't take any photos of Da in SCK during my visits since we always thought there would be plenty of other times for that.

Da Ruan's office in SCK. and the model of his development of fuzzy logic in nuclear reactor

"*Huo Lei Feng*" (活雷锋)

I don't know how many times Da had been a referee for his colleagues and friends. He didn't list it in his CV, but many friends have told me that he refereed for them, some of them several times. And I know Da didn't know how to say "No" to his colleagues and friends, if he could help; I know Da was always approachable. He kept checking emails even in his holidays. His mum told me, when I visited her during ISKE2011, that when he visited Shanghai, he always tried to check emails in an Internet-cafe while his parents were resting.

Da was always happy when he was giving help to others. He told me a few times that he wanted his actions to resemble those of Lei Feng (Note: Lei Feng is a Chinese hero who helped many people in his short life.).

A humorous man

Da had his own brand of humour. He loved to tell hilarious jokes. One is about pizza, which I heard several times on different occasions.

> "*When Marco Polo visited China about a thousand years ago, he saw a kind of traditional Chinese food called Bingzi (only Shanghai people call it Bingzi, and Da had a Shanghai accent). He asked his assistants to learn how to make it. After they returned to Italy, his assistants forgot how to put the meat and vegetables inside and just put them on top of the base. They also forgot the correct pronunciation, so Bingzi became Pizza.*"

Because Da looked serious whenever he told this joke, every time there would be people who asked, "Really?" which always made me laugh a lot..

Continuing Da's work

After Da's passing, people in our community found various ways to channel their grief for their close friend. In the first couple of weeks, I received over 100 emails which spread the tragic news and expressed people's experiences with Da throughout the world, in memory of him. The large number of emails was testimony to the high respect in which Da was held, and how widely he was liked, both personally and professionally. All the emails were full of accolades, admiration, high praise and honour for Da Ruan.

After a few weeks, I started to receive many emails from authors, editors, publishers, conference organisers, guest editors, his students and post doctoral candidates, all wanting to discuss how Da's incomplete work should be continued. He had at least five unfinished papers or papers under revision, draft proposals, two agreed special issues as a guest editor and seven special issues under his organisation for the IJCIS; two confirmed conference keynote speeches; two conference chairs; and two book contracts needing his signature, as well as his job as editor-in-chief for IJCIS and editor-in-chief for a book series. More importantly, he was the chair for three conferences: ISKE2011, FLINS2012 and SSRAOC2012. I knew Da had been working on many research projects, led different groups and communities, and collaborated with many institutes, but I could not imagine what a high workload he took on, and how many plans he promised to the community.

Dear Da, your energy, your unlimited devotion, and your sincere enthusiasm will encourage us forever. We must continue your work by providing a friendly, compassionate community to live up to your incredible standards.

It was very hard for me to edit this book and also hard to write this chapter. Da still stays in my heart. Sometimes, even while I was writing this contribution or editing this book, I was still thinking of sending him an email. When I attended ISKE2011, I even thought I would meet him at the conference. When I realise he is in another world, my heart feels great pain. We have lost a great and very genuine friend. We lost a community leader. We lost our brother.

Nicolas said that his Dad is travelling. I believe he is right. Alexander wrote very nice music and we called it "Peaceful Garden". We wish Da a good trip in the peaceful garden.

Guangquan's Memories

I have known Da for 15 years. In December 1996, I was invited by Prof. Kerre to visit Ghent University for research cooperation. I shared the same office with Xuzhu Wang who was Prof. Kerre's PhD student. One day morning, Xuzhu called me "Please come to Prof. Kerre's office, Da Ruan is here, and he is Prof. Kerre's first Chinese PhD student". This was the first time I met Da. He impressed me with his open-minded conversation, though our meeting was very short. In October 1997, I visited Prof. Kerre for a second time and met Da again at Ghent University. He told me about his project in SCK and his FLINS conference but the meeting was still brief. At that time, I had just moved to Perth from China, and I then moved to Sydney, so we lost connection until March 2003 we met in the conference "Fuzzy information processing: theories and applications" which was held in Beijing. We had a longer conversation this time. He invited me to attend a special session, organised by Prof. Kerre and himself, within the Joint Conference of Information Sciences. We then met in New York in 2004. We had more time then to discuss our research interests and established a close research cooperation, becoming best cooperators, best friends and best brothers. We co-authored two books, 3 book chapters, 8 journal papers, 6 conference papers, and three special issues.

Da in his UTS office in 2006

In December 2005 and November 2006, Da was invited to visit our lab at the University of Technology, Sydney. In 2005, he attended AI2005 conference and gave a whole day tutorial on "Tips for implementing applied AI in complex systems". He gave many real word examples of complex systems including his nuclear projects to demonstrate the implementation of AI in the real world. Many people expressed their panegyrics about Da's brilliant tutorial. In his visit in November 2006, he gave an important and attractive seminar, “How to Write High Quality Journal Papers” to PhD students and staff in our university, and received high appreciation. Five years later, many people still remember this presentation and have followed his advice in writing papers with success.

One weekend in November 2006, during his visit to Sydney, we invited him to spend a weekend at my house. We bought some fresh and tasty seafood. I opened a bottle of famous Chinese Wine, Moutai, with 48 degree, which had been kept for many years. We were eating, drinking, and talking, very excited and happy, until 3am. One bottle of Moutai was finished—Da drank more than half of the bottle. He really liked Moutai and was a good drinker! I gave him two bottles as a gift after that. I particularly remember that we discussed our research grants, joint papers, and special issues of international journals that night, and especially, we talked about his initiatory plan to start a new journal in computational intelligence applications. He launched this journal, the International Journal of Computational Intelligence Systems (IJCIS), in 2007 and its first volume was published in

January 2008. The first step is the hardest one. This journal is doing well and its reputation has increased a lot, but Da can only look after this journal in another world.

Da had a wonderful time in Sydney during his two visits. See the kind email below that he wrote in Chinese in Dec. 2006. When I re-read it, his voice and expression are still very fresh before my eyes. He also planned to visit Sydney at the end of 2011 with his family, but now he is travelling in another world.

路节，广全及张最：
衷心感谢您们对我的热情款待。在UTS的日子里，我感到象一个皇帝一样。我所收到的不只是悉尼城的世界级美丽，UTS的优美环境，还有很多很多在MOL无法获得的中国城的佳肴。路节的超一流水饺，汤米，烙饼，花生米加上广全的茅台，还有和张最一起用的午茶。当然，还有其他所有的好吃好玩的，都让我铭刻在心。特别是第一次品尝到鲍鱼，生猛海鲜，让我不再要美味的粽子。虽分不清"虾"与"蟹"，但吃得可乐了！如果说悉尼比作是大海，那么MOL只能是大海边上的一颗沙子。悉尼之游，可谓永世难忘！
顺祝圣诞及新年快乐！
达

We met almost every year in recent years at conferences: in Xiamen for ISKE2008, in Hasselt for ISKE2009, in San Antonio in 2009 for the IEEE International Conference on Systems, Man, and Cybernetics, in Chengdu for FLINS2010 … we always talked about that weekend in my home and he always told me some of his new plans.

Da was not only an excellent academic and a nice friend and a leading scholar, he was also a wonderful father. I shall always remember how proud he was when he talked to us about his two sons, Alexander and Nicolas; his eyes were filled with the love of a father. I particularly remember how hard he tried to shop for nice gifts for his sons, in New York, in Sydney, in San Antonio … whenever he was travelling. He loved to spent time searching, selecting interesting gifts to hid boys. He told me many stories of his two boys, from their study to their sports; from the food they liked to the games they played on the Internet.

Da passing away so suddenly left me not only with deep sadness but also with endless recollections of the stories from the past. Since 1 Aug 2011, I have recalled many memories of Da. His kind voice and countenance as well as his important contribution to our community will stay with us forever.

We are proud to have this friend and feel lucky to have worked with him.

Da, my good brother, you will remain in our hearts forever.

Da and Guangquan in Sydney, Nov. 2006

Jun's Memories

August 1st, 2011 would have been an ordinary day in my life but for an unexpected call. "Check your email! Prof. Da Ruan passed away yesterday!" My wife's hurried voice came from the other end of the line. When I opened my email, a message from Prof. Jie Lu confirmed the bad news. It could not be—however, it was true. Our best friend and my strict supervisor, Prof. Da Ruan, had gone to Heaven for ever.

In the following week and with deep sorrow, Prof. Jie Lu, Prof. Guangquan Zhang and I organised a memorial photo album on behalf of Da's Chinese friends. Although I tried hard to pretend that we were doing it for someone else, I could not ignore Da's smile on those photos. I had planned to find a photo with him but, regretfully, I could not and it seems there is not one, even though almost 12 years have passed since I first met him. His everlasting smile brought to my mind the recollections of past years.

In 2006, Da spent a couple of weeks in Sydney for collaboration with Prof. Jie Lu and Prof. Guangquan Zhang. One day in the car home, we recalled the SCK dormitory where we have all experienced good times on different occasions, the beautiful downtown of Mol where he worked and lived until the last day, and the food he made as our host. Suddenly, Jie asked me, "Da always boasts that the 'white-cut chicken' he cooks is the best! What do you think? Tell me the truth."

"Very good, but I am not sure if it's the best!" I said seriously. "Really? Why?" Jie asked again. "Because, you know, Mol is too small, I have nothing to eat and I … (I had never compared it with others')!" "Ha! Ha!" Jie, Guangquan and Da laughed heartily before I could finish. I tried to add some explanatory words. They laughed even more. Mol is a very small town in Belgium, and the only visitors to Mol from China are researchers from universities or research institutes. Da liked to cook for every visitor, I guess, because he tried to reduce each visitor's feeling of being alone in a faraway and strange place. In fact, the chicken he cooked is really the most delicious food in the world. It reminded us of home.

Da enjoyed life and family to the full. In 2003, I worked in Mol under his supervision for a few weeks. One day, he invited me for lunch in his home. When I arrived, he was preparing Chinese food. He shared his secret recipes with me and taught me how to make a Shanghai food using eggs and vinegar. After lunch, we chatted about interesting things and drank Chinese tea while his two children played in the room. He watched the two boys' horseplay with absolute love in his eyes and recounted many funny stories of his sons, such as how Nicolas, his youngest son, won free drinks at a football club. He told me that being a father is a hard work. Many years later, I can still remember the love in his eyes. After Professor Jie came back from Da's funeral ceremony, she told me something that Nicolas had said: "I pretend in the morning that my father has gone to travelling for work, but in the night, I know he will not be back any more." Hearing this, no words could express my feeling.

Da was a strict and diligent scientist. I first met him in March 1999 in Chengdu, when he was visiting our university with his family. One day, I went to his hotel to collect him and his family for a scheduled tour to EMei Mountain. When I arrived at his room, he was working at his laptop and beside him was a pile of plastic slides. He asked for my help to connect to the Internet because he wanted to verify some data for his lecture the next day. After the hotel reception helped to open the network connection, he thanked me and told me that he wanted to stay at the hotel to prepare the next day's lecture. I tried to persuade him to take the tour, but he insisted on completing the slides first. Finally, I gave up. Since 1999, he has visited Chengdu many times to attend conferences and deliver lectures, but he did not have his first and last visit to EMei Mountain until 2010.

Da was my respected supervisor and friend. He always frankly and pertinently expressed his opinion. One day in 2007, I received a long email from him. In it, he made a critical analysis of my research work and made some suggestions. I could feel a friend's deep concern behind his words. I printed this email and pinned it to the notice board in my office. It always woke me up when I felt too lazy to work. We collaborated for many years in China, Belgium, and Australia, and he always tried his best to help me when I needed it. In 2003, he invited me to work at Mol for a brief term, which gave me the first taste of a different research environment and culture. In 2006, he recommended that I should work with Prof. Jie Lu and Prof. Guangquan Zhang in Australia, which expanded my research experience. In 2009, he accepted Professor Jie and Professor Guangquan's invitation to be my external supervisor to support my future career. However, since August 2011, I will never again have the chance to say “Thanks” to him.

Da left; however, he left us endless smiles and encouragement to face every new day. Da, we will always remember you! R.I.P.

Fengjie’s Memories

It was a pleasant experience working with Prof. Dr. Da Ruan in co-authoring the book “Multi-Objective Group Decision Making: Methods Software and Applications with Fuzzy Set Techniques” in 2006-2007. In 2006, Prof. Da Ruan visited the Faculty of Information Technology at the University of Technology, Sydney (UTS) for three weeks to finalise the book. I was privileged to have the opportunity to work alongside him.

What impressed me most was his professionalism and work ethic. His quick response to my inquiries and his coordination of me and other authors made our writing, software programming and testing much smoother and less stressful. He always listened patiently to my ideas and opinions, and adapted any reasonable suggestions. Due to the heavy workload and tight timeframe, he had to work long hours every day during his stay in UTS, yet he never looked tired. I was amazed by his enthusiasm, passion and energy for this book and his work. His attention to detail was not only on the content and structure, but also on the wording, grammar and punctuation through the whole book. This was ultimately a practical demonstration of his seminar at UTS in 2006 entitled "How to Write High Quality Journal Papers". This experience has benefited me extensively since then.

In spite of his highly recognised achievements, he was still very humble, caring and easygoing. I still remember that he brought Belgian chocolates for every member in my lab when he came to Sydney.

I still cannot believe he has left us forever, and I am saddened by his sudden passing. RIP Prof. Da Ruan.

Chao’s Memories

I met Prof. Dr. Da Ruan when I was still a PhD student under the supervision of Prof. Jie Lu. I remember it was around the second year of my PhD study. At that time, my supervisor had already helped me a lot with my PhD research in terms of research methodologies and study directions. I started to write my research proposals and build the framework. However, I was still not fully confident about what I would build. It happened that Prof. Da Ruan came to visit our Lab and Prof. Jie Lu introduced him to us. It was the first time I met Professor Da Ruan and I was very impressed by his high spirits in doing research.

I briefly talked about what I wanted to do and he listened carefully and also asked a few questions to clarify his understanding. He then gave his advice and encouraged me to go further. With the additional encouragement from such a senior researcher, my confidence was boosted. His cheerful and energetic way of doing research also influenced me, making me feel more inspired in pursuing my research direction. It is pity we didn't have much chance to meet with each other again, but I always have the impression of him as a cheerful and encouraging researcher.

Jack's Memories

It was in 2006 that I first met Prof. Da Ruan. At that time, I was visiting UTS for one year. He was a perceptive and quick-witted scholar and he was dynamic and energetic when he focused on his research. His talk "How to Write High Quality Journal Papers" made a great impression on me and provided us with valuable knowledge and better ways of thinking ways. I met him second time at UTS when I had enrolled as a PhD candidate. In 2009, I met Prof. Da Ruan for the last time in Beijing for the RACR 2009 Conference. We shared discussions after conference sessions and he gave me many good suggestions for my thesis topic. My research had been successfully conducted and completed, and I was lucky that he accepted being one of my PhD thesis examiners. His comments were constructive but challenging. Not long after I received his comments, I heard the sad news. It's unbelievable that we have lost such a great person.

Da gave a seminar at the University of Technology Sydney in Nov. 2006

Julwan's Memories

I came to know Prof. Da Ruan in 2009 when I planned to enrol as a PhD student at the University of Technology, Sydney, Australia. Prof. Jie Lu, my principal supervisor, introduced me to him. Since then, we have communicated by email and talked about my PhD research topic. He was really kind and gave me invaluable direction for my research. Together with Prof. Jie Lu and Prof. Guangquan Zhang, we developed a new approach for assessing nuclear power plant safety. We proposed a fuzzy reliability approach as an alternative to a probabilistic reliability approach to evaluate nuclear event failure rates when reliable historical data is inadequate or unavailable for statistical calculation. Our collaborations were fruitful and in that time we co-authored a number of papers related to my PhD research project.

In 2010, we met at the FLINS conference for the first time. I really enjoyed talking with him. We discussed the future of my research and again he made invaluable suggestions. At that time, he promised to visit me at UTS. Sadly, he has gone to the heavenly Father forever before he could realize his promise.

To me, Prof. Da Ruan was not only an outstanding researcher and a great supervisor, but also a warm-hearted friend. During our meeting and discussion at FLINS2010, I could see his leadership and I am really honoured to have had him as my supervisor. Rest in peace Prof. Da Ruan.

Prof. Dr. Da Ruan and Julwan at the FLINS 2010 in Chengdu, China.

3.6 Miss You Forever, Da Ruan, Consulting Professor at Southwest Jiaotong University since 2000

Southwest Jiaotong University, Chengdu, China

An internationally highly recognized expert in computational intelligence and decision making and senior researcher of Belgian Nuclear Research Centre (SCK•CEN), Belgian-Chinese Professor Da Ruan died unexpectedly on July 31, 2011. We are deeply shocked and grieved for losing an outstanding consulting professor.

Professor Da Ruan was employed as a consulting professor of our university since 2000. Over the years, the unit that Professor Ruan had the most academic activities in China was our university. He tied to our university and made one even more special trips to our university every year. Professor Da Ruan had made tremendous contribution to international cooperation, external communication, personnel training, scientific research of our university.

In the field of international cooperation, Professor Da Ruan helped School of Mathematics, School of Information Science and Technology, School of Transportation and Logistics of our university to develop cooperation relationships with colleges and research institutions abroad actively. Under his effort and assistance, we created close relationships with some colleges and academic organizations in Belgium, France, United Kingdom, Germany, Australia, Spain, Japan, America, Greece and other countries. It greatly contributes to the extension of international cooperation of our university.

In the field of external communication, Professor Da Ruan helped our university to organize a number of international conferences and made great contribution to the international academic exchange of our university. The second International Conference on Intelligent Systems and Knowledge Engineering (ISKE2007) was held in our university in October 2007. He gave careful guidance to each preparatory work of the conference and revised every document carefully from the letter of call for papers to the program of the conference many times during the preparatory process of the conference via hundreds of e-mails and nearly one hundred phone calls. He also attended the conference and gave us specific guidance and assistance as a general chair of ISKE2007. The ninth International Conference of Fuzzy Logic and Intelligent technologies in Nuclear Science (FLINS2010) was hosted by our university in August 2010. He gave us detailed guidance on the agenda of the conference. He attended the conference as one of the general chairs of FLINS2010. He also invited many world famous experts on intelligence information processing to attend the conference. This conference attracted 111 experts from 15 countries, among which there are 64 oversea attendants. He played an important role in the promotion of academic exchange in the field of international computational intelligence. What's more, the first International Symposium on Massive Data Mining and Knowledge Discovery (MDMKD2009) was initiated and organized by our university in September 2009. He also gave support energetically and delivered a keynote in his busy schedule. He also supplied a large amount of information for teachers and students of our university to participate in various kinds of international academic exchanges and gave continuing support and assistance.

In the field of personnel training, Professor Da Ruan took the utmost care to guide more than 20 PhD students of our university. He often went back and forth between Belgium and China at the expense of holiday time tirelessly and gave academic reports, guided students on writing and attended defenses or academic symposium. He also often tutored students on their research work and helped them revise papers, and provided academic information through e-mails specially. At the same time, he supplied opportunities for several students to study and exchange overseas. He made outstanding contribution to personnel training of our university.

In the field of scientific research cooperation, Professor Da Ruan kept close academic cooperation relationships with many professors and many schools of our university since 1997. Under the cooperation relationship, we undertook many international scientific and technological cooperation projects and published two monographs, six international conference proceedings and more than 60 papers. He made distinguished contribution to improve the scientific research level and extended the field of international cooperation of our university.

What Professor Da Ruan left to us is not only limited to the contribution to the development of our university, but also precious spiritual wealth he left to us. "5.12 Wenchuan Earthquake" brought huge disaster to many people, and also affected the heart of Professor Da Ruan. He made a special trip to our university to visit teachers and students just few days after the earthquake. His family also launched many European friends to donate goods for the area of earthquake. His two children under 10 years old produced special PPT files to introduce the situation of the earthquake zone in their school and appealed other children to offer compassion for the area of earthquake. The pure heart and innocence love of Professor Da Ruan worth learning forever.

Even through Professor Ruan had left us forever, the great contribution he had made to our university and the spiritual wealth he had left to us will never left from the history of our university forever. Professor Da Ruan will long live in our heart forever!

We will remember our excellent consulting professor, Professor Da Ruan, forever!

3.7 Professor Da Ruan, Distinguished Visitor at Complutense University, Madrid

Prof. Dr. Javier Montero[1], Prof. Dr. Begoña Vitoriano[1], and Prof. Dr. Daniel Gómez[2]

[1] Facultad de Matemáticas
[2] Escuela de Estadística, Universidad Complutense de Madrid, Madrid, Spain

This short note is written as a contribution to the memory of Prof. Da Ruan. We shall try to reflect how he was entitled to stay two months (April-May 2011) as "Distinguisher Visitor" at Complutense University, Madrid, based upon his past collaborations without team, and pointing out some of the projects we jointly developed and launched during such a period. Beside the strictly scientific arguments towards his academic excellence, we want to stress his continuous support to explore new initiatives from a sincere and friendly collaboration.

Complutense University is the biggest University of Spain in number of regular students (i.e., excluding those students from on-line courses). More than 80.000 students and almost 8.000 professors join Complutense University. The programme for Distinguished Visitors was created some years ago to bring to our university leading researchers that could help to launch new strategic projects. In order to be selected as Distinguished Visitor at Complutense University it is taken into account not only excellence in research, but leadership and ability to give in few months a definitive impulse to some strategic project. Professor Da Ruan got such a Distinguished Visitor ship for the course 2010-2011, and spent in our university two months, from first April to first June, 2011. During those two months in Madrid, Da Ruan worked within our Decision Aid Systems research group into two main projects.

On one hand, Da Ruan worked with our aggregation working group, led by D. Gómez, studying the properties of consistency and stability of aggregation operators and obtaining some interesting mathematical results with potential implications in several fields. Some preliminary results have been published in several conference proceedings, and put into a paper co-authored by Da Ruan, still to be published. In this paper we have shown the necessity of considering some consistency and stability definitions within some families of aggregation operators. Within this framework, Da Ruan also worked with us in the design of a COST European Project entitled "Integrated Uncertainty Analysis Theories for Complex Systems", which will be hopefully submitted in January 2012. The main aim of this project is to integrate various uncertainty analysis theories such as Probability Theory, Fuzzy Sets Theory, Intuitionistic Fuzzy Sets, Rough Sets, Data Mining, Possibility Theory, Lattice Valued Logic and Evidence Theory into a unique general mathematical framework that can deal with different uncertainty-based information such as incomplete, imprecise, fragmentary, not fully reliable, vague, contradictory and overloading in complex systems.

On the other hand, three additional main achievements specifically focussed on disaster management were launched during those two months of Da Ruan while in Madrid: a joint paper on a new continuous extension of Belnap's four valued logic, an Atlantis Press-Springer book on decision analysis and support in disaster management to be edited by J. Montero, D. Ruan and B. Vitoriano, and an ambitious project recently submitted for funding within the European Union framework. This Disasters Management working group is led by B. Vitoriano, which was the nominal hostess of Da Ruan as Distinguished Visitor at Complutense University. Da gave this working group a view of the future, and accelerated (even more) our rhythm of research, launching us to risky and exciting academic adventures. Da gave us organization skills and strength based upon an intense and extraordinary daily academic activity, still allowing room to deep and quite conversations about less technical issues, either professional or nonprofessional. Above all, he gave members of this working group confidence in themselves, confidence to tackle challenges. His visit marks indeed a before and after in our Disasters Management working group. We were exceptionally lucky to have him among us. Da pushed this working group to request all previous and establish new strategic contacts in disaster management and closely related research fields, which has been finally fruitful in a consortium which has already began to act as a team.

All these experiences will prove to become steady collaboration projects, as Da made clear in the meetings he organized in Madrid, where a number of prestigious researchers were enthusiastically involved in some of the above projects (another meeting had been scheduled in Malaga last September, and we had to sadly meet missing his leadership). Not having Da Ruan with us any more is indeed painful for us, and we have needed some time to recover from such a terrible shock. But we will make our best to continue his work and fulfil his vision. His work will be fruitful beyond all the above particular projects, simply because they will bring all its participants into joint long term projects, and such collaboration will bring highest objectives to be pursued. His departure left us puzzled, but the vision he brought to us will make us continue and develop all the projects he pushed with us, and that we could not have started without him. Yes, there is a before and after, thanks to Da this Disasters Management working group is greatly expanding, and we look forward to future attractive objectives.

Beyond our research group, Da impressed our colleagues in his several talks at our Faculty seminars, but also in his daily affability and proximity. Da continuously showed how to keep under control a number of variables, still creating a friendly environment from the first day, with no stress despite the overloading of work and meetings. A lot of researchers at Complutense University got benefits from Da Ruan's visit. Thanks Da!

Some Final Words

Sadness is difficult to avoid when an unexpected loss of a friend and colleague like Da Ruan happens. As his colleagues, we shall remind his outstanding work, his endless work and his devoted dedication to promote and support joint scientific initiatives. But as friends we shall keep in our mind lots of memories, his sincere friendship and his smile. We are proud that Da considered us his friends, and we are also proud of having the opportunity to share the above projects with him. Quite often our own busy life makes it difficult to realize to how much some people are influencing us, how relevant they are for our own life, how much they deserve to be admired as a human being. In this sense, it might be the case that Da never realized the extent of his influence in all those projects around him, and the deep gratitude many of us owe to him. The only sure thing is that Da was proud of his family. Da's life will definitively illuminate his beloved wife Greet and his beloved children Alexander and Nicolas, but also all those who pretend to continue his projects and took advantage of his wisdom. Our research team on Decision Aid Systems is one of them.

One of the authors would like to express his personal gratitude and admiration to Da's wife, Greet, who from her terrible moments helped him so much when he travelled from Spain to Mol, just to say goodbye to Da at the funerarium, a very sad trip.

Da, Jie and Javier in Plaza Major, Madrid, Spain, Jan. 2008

3.8 Memories in Complutense Campus

Prof. Dr. Victoria López, Prof. Dr. Matilde Santos, Prof. Dr. Guadalupe Miñana, Prof. Dr. Raquel Caro, Susanne Escobar-Torres Kraemmer, Prof. Dr. M. Sánchez Balmaseda, and Prof. Dr. Guillermo Botella

Dep. Computer Architecture and Automatics, Faculty of Informatics, Complutense University de Madrid, Madrid, Spain

This chapter is a summary of the most relevant activities that Prof. Da Ruan carried out at the Faculty of Informatics, University Complutense of Madrid during the last three months of his life. His work covered many areas: research, teaching and social relations. In all of them he was a great professional. With this chapter we want to pay tribute to him and thank his friendship forever.

Victoria's Memories

Da and I met first time in Jaén, a nice city in the South of Spain, in 2004. He was invited as main speaker at the Spanish Fuzzy Conference. His lecture on "*Fuzzy techniques in control and decision making for complex system*" was really brilliant. I have to say that his personality impressed me from the beginning. He also attended my oral presentation even it was in Spanish. I still remember him smiling at the end of the room. Luckily, we were able to spend much time together and even to go shopping in the old part of this beautiful city. Since then, every year we met at several international conferences and he always introduced me to very interesting people. I say this because a lot of the people I know have been and are in fact, personal friends of Da, and all of us know that his personality has opened many doors. I really feel very special when reflecting and remembering how Da has directed my personal and professional life during the last years. I have special memories of Genoa in 2006 where FLINS conference was held, mostly because of the proposal of Madrid as the new host city for next FLINS conference. FLINS 2008 in Madrid was so special for a lot of good advices he gave me in a personal conversation that I will never forget. I have also special memories in Xiamen, Barcelona, Madrid, Hasselt, Chengdu and Salt Lake City.

Stay in Madrid 4 April – 4 June, 2011

On April 4, 2011, Da Ruan arrived to Madrid for a stay of two months. Prof. Matilde Santos and I proposed him to come one year before and he accepted almost immediately.

Da Ruan and Victoria López in FLINS 2008

On April 5, we had our first meeting at Mathematics building. We were in the coffee break when Prof. José Carrillo appeared and I had the opportunity to introduce him to Da. At that time Prof. Carrillo had put himself forward for chancellor. Since the first round of elections was the next day, Prof. Carrillo invited us to follow the event of the votes recount. Da was happy. Since then, they also became friends in some way. The process was very interesting and exciting, especially because Prof. Carrillo finally won the first round.

We spent a lot of time together going to the cinema, concerts, and having lunch and dinner with other friends, but also for making some research work about decision making and lattice algebras. He encouraged me to work on this last topic since many colleagues from China were making a successful work about, so we started to prepare a new paper to be sent to FLINS 2012. Besides he was very worried about the 1st International Workshop on Safety & Security Risk Assessment and Organizational Cultures to be held at Antwerp, and to which several papers written by some colleagues and me had been sent at the time. Finally some of them have been selected for oral presentation.

On May 8, we both travelled to Jaén by car. He drove while he was speaking proudly about the academic progress of his sons, especially of Nicolas at Mathematics. We spent this time not only for leisure but also to lay the foundations of future agreements. In Jaén, his good friend Prof. Luis Martínez was waiting for us and we all enjoyed a really good time visiting the new Research Center in which Luis was the header in fact.

On May 13, we also celebrated together the second (and final) round of elections. Prof. Carrillo won again and gave us two invitations for a special dinner at Siglo XXI Society, a very high level Cultural Society in Spain, on May 17. Da enjoyed specially that dinner. He was talking happily with all the people, many of them members of the new team government of the UCM. All of them still remember him.

On May 19, he gave us two speeches at Faculty of Informatics. The first one was entitled '*Elementary guidelines for preparing a journal article in Computer Science and Engineering*'. This talk was within the program of Cátedra Miguel de Guzmán and NovelMat Project of Complutense University, in which I was working as an active member. A lot of students and young researchers attended it. Some of them shared his good advices and tricks by Facebook and Twitter after that. The second talk was a researching one: '*Uncertain information analysis and intelligent decision support in perception based complex systems*'. This time a group of teachers and senior researchers had the privilege of attending it.

On June 3, the inaugural of new Rector Magnificus was held. Da joined us to attend and celebrate the ceremony, as his last event in Madrid. He left Madrid on the following day.

More about Mol and Shanghai. June, 4- July, 4

Even from the distance, one could notice the great working activity of Da. On June 10, he traveled to Shanghai where he spent the Sunday at his parents'. This was the last time they saw him alive. He came back to Belgium on June 16 from Beijing. Again, he was always ready for new proposals and agreements. He accepted to be a teacher in my summer courses next July. On June 17, Da also signed a document addressed to the Ministry of Education of Spain to become a professor of Computer Research Master in my course in Bioinformatics. He wanted to prepare some lectures about Data Mining and Decision Making.

On June 20, Da sent a letter to Prof. Carrillo in which he said:

'(…) please accept my sincerely congratulations on your being the Rector at the most prestige UCM in Spain. (…) It will be my honor and pleasure to bridge UCM and any of the Chinese universities for any future cooperation. (…)'

At this point I want to say that Da used to send me a lot of documents. All of them are very carefully stored since then (and available for anyone on request).

Last Visit to Madrid 4 July – 6 July, 2011

He referred to this short visit as his *compact agenda*. On July 4 early morning, I sent him a document ready to be applied to 2011 International Prize Cajastur on Soft Computing by European Center for Soft Computing (ECSC). He had asked me to help to him organize several documents. This prize would mean an increasing in the prestige of the International FLINS Conference, where he was the General Chair. I remember him very excited with this issue. I sent him the final document at about 7.30 a.m. and he replied me with everything done by 8.15 a.m. This same day he came back to Madrid. We met in the afternoon at Madrid-Barajas airport. He was plenty of life and very happy.

On July 5, we met at 7.30 a.m. for attending a conference and some meetings at CDTI (Centre for Industrial Technological Development of the Spanish Ministry for Science and Innovation). We joined there a group of people from Complutense University of Madrid, University of Jaén, Pontificia Comillas University, and Polytechnic University of Madrid. In fact, Prof. Da Ruan had kindly accepted to be leader of a proposal for the European Project (VII Frame Program - Cooperation Work Program 2012 "Development of resource efficiency indicators"). All of us enjoyed a funny lunch time. Then, we were very impressed by his brilliant speech with the officials of CDTI. Da had many skills, one of them was gift of speech. He knew about which word fits at any time, at any situation, and that afternoon all of us stayed in silence while he made a wonderful explanation of the proposal.

After that busy agenda, we met again at 7 pm for attending the inauguration of the UCM Summer Course. The ceremony was driven by the vice-chancellor Prof. Carrillo and Rosa María Calaf, a famous journalist who was invited as plenary speaker. The talk was about the vision of foreigners about the Spanish people, and Rosa Maria talked about a book that Chinese children used to learn Spanish traditions. By then, Da understood Spanish well enough to enjoy the conference and we were having really fun due to amazing examples of misunderstandings within the book of Rosa María. Finally, during the cocktail venue, Da and Rosa María, were talking animatedly until quite late.

On July 6 early morning, he participated as a visiting professor in my Software Engineering summer course with two talks about Business Intelligence. I want to thank the 48 international students who attended his talk, their messages of condolence and words of appreciation to Prof. Da Ruan. Also I'd like to give thanks to all of them for the pictures they took of his speeches, which are the latest images that I have of him. Before the meal, we attended an important meeting with Juan Ferrera, the new vice-rector of International Relations of Complutense University. Da had offered himself to establish agreements between the UCM and several Chinese universities. The meeting was very fruitful and Juan Ferrera was greatly pleased with his cooperation. Finally, we went to El Pardo for lunch, a beautiful small city nearby. We were joking about his *compact agenda*, and making a funny review of all our work together. I have in my memory the image of his face and smile at the airport when he told me goodbye with one of his winks. That was the last time I saw him alive.

Da teaching Business Intelligence at UCM in July 2011

The End, 7th to 29th July

Since he returned to Mol, we were in contact via Skype and email in order to make progress in some issues. One of them was the cooperation agreements with China and the meetings I should attend in his name at UCM. Other issues were about helping in preparing my special session on Formal Engineering and Reliability for ISKE 2011 and the proposal for a book on this subject, in which a group of colleagues of mine would participate as editors. The hardest work was probably the proposal of the initial document for the VII Framework Program and its organization. I still have about 50 emails he sent me during the month of July, and a total of about 250 during the last year. I want to quote here a fragment of one of them dated on July 29th, 2011. With this mail I want to show how Da was working at his holidays, and how he was always worried about friends, family and work.

'(…) I'm working on a few book chapters (…).

After Aug. 1, I'll have a lot to do for SCK. One project on nuclear safety culture with a new postdoc under some financial support from some Belgian nuclear company (will have a meeting with that nuclear company on Aug. 9); the other project on nuclear safeguards open information analysis with IAEA (Int. Atomic Energy Agency in Vienne) (will have another meeting with IAEA people on Aug. 11 at SCK). I just need to prepare some research plan for them and hope to get some support from them as well.

Now Tianrui Li will send one of his PhD students to SCK for one month (From Aug. 1 to Sept. 2), (…). He will be here next Monday morning at 8 am, but I have arranged a taxi to pick him up in Brussels. I'm also a co-supervisor for his PhD.

In addition, I work on IJCIS where I'm the Editor-in-chief. Now the journal (2010) got its impact factor of 1,47 (vs. 0,31 in 2009 for the 1st time). This is my main task during the summer vacation. I could read most submissions right now. During the normal working days, I don't have much time to take care of the journal. (…)

Take care,

Da'

After his passing away, with a great pain in my heart, I could check all emails in my box, text messages in my mobile, historic chatting files in my Skype, … I have a lot of documents from him, even his Chinese e-signature, that he gave to me in case of necessity. I have got an email box full of condolence messages as well.

All of us have lost a very good friend. Always in my heart.

Some Friends' Memories

'I had the privilege to meet Da and to consider him a friend of mine. Being as he was a brilliant researcher and academic, he did not mind helping people who were starting the academic career. He devoted his time to helping other with his guidance. Everything seemed easier with him. He made you work hard by proposing challenges than eventually benefited you. In fact, I even did things that I had never thought of, like dancing *sevillanas* in the gala dinner of the International Conference FLINS 2010.

I was also lucky to see each other attending some conferences, many of them organized by him. He was not only aware of the technical success of the scientific program but taking care of us and making us to enjoy the different activities. When he was in Madrid, we had lunch together several times, and he always took interest in you. I would like to keep such nice memories forever.'

Prof. Dr. Matilde Santos
Complutense University of Madrid

'I met Da in Hasselt (ISKE 2009) where Matilde, Victoria and I were presenting a research work. I also had the opportunity to work with him during his stay in Madrid in spring 2011. From the first time, he showed interest in our work and gave us full support, even helped us with the name of our research group.

It is not easy to find people with such human and professional qualities as Da. For me, it is a privilege to have known him.'

Prof. Dr. Guadalupe Miñana
Complutense University of Madrid

'My very special thanks to Professor Da Ruan for not only giving me the opportunity and the great privilege to discuss my work with him respect to VII Framework Program (in which he kindly accepted to be the leader) but also for his keen intuition, his guidance, his compendious advice, his valuable suggestions for the improvement of the proposal and his rigorous scientific professionalism. I will remember our colleague as a dedicated and inspired teacher with nice personality. In July I enjoyed his stay in Madrid with Victoria, María José and Javier mainly because of the espaguetti frutti di mare at Nicoletta! Da, we miss you!'

Prof. Dr. Raquel Caro Carretero
Pontificia Comillas University, Madrid

'My good friend Victoria introduced me to Da Ruan the day I invited her to a special concert given in the beautiful Art Deco Palace of Longoria in Madrid, promoted by the department of cultural cooperation of the Austrian Embassy and with assistance of the Ambassador. I was impressed by the easy, direct and warm personality of Da. I often remember the conference he gave at the Universidad Complutense de Madrid about guidelines for research articles, transmitting such clearness, energy and straight forward thinking with his words. He was one of these rare, intelligent and interesting people it's a privilege to meet in live.'

Susanne Escobar-Torres Kraemmer
Austrian Embassy of Madrid

I never worked with Da Ruan, unfortunately, and, unfortunately again, I never listened to one of his lectures. But I was lucky to share with him some of the intense moments of the election process of our chancellor, moments of expectation, happiness and fun, and I remember him, always smiling and taking part in it as anyone. The last time I saw Da, having together a snack after the

UCM Summer Course inauguration, I asked him how he managed to be in all the university events and parties of the last months. "It is Victoria´s merit. She always takes care of me and brings me to the best places like this. That's why I always come back to Madrid", he said with his generous smile.

Margarita Sánchez Balmaseda
Complutense University of Madrid

'Da's enthusiasm and mate ship will be fondly remembered by me. I will miss his passion for his research, his students, his colleagues, the partnerships he established, and for his many projects. I will miss his sense of humor and his visionary leadership. Still I remember his great talk about gold rules communicating scientific ideas.'

Guillermo Botella
Complutense University of Madrid

3.9 A Fresh Breeze of Excellence in Jaén

Prof. Dr. L. Martínez, Dr. M. Espinilla, and Dr. R.M. Rodríguez

Sinbad[2] Research Group, University of Jaén, Spain

In this chapter we want to express not only our friendship, gratitude, appreciation and thankfulness to our respected friend Da, but also to highlight his support and cooperation since 2004, first time he visited our city and university. His help has been a key for our research careers both in knowledge and research opportunities. Hopefully this chapter can help to pay a deserved tribute to him.

Luis' Memories

I am still remember first time Da visited Jaén (small city in Andalucía, South of Spain), after two short meetings in 2003 and 2004 I invited him to be plenary speaker in the Spanish Fuzzy Conference that was held in Jaén in the middle of September 2004. He arrived to Jaén on Tuesday 14th September and as usual it was a hot day over 30 degrees. His talk on *Fuzzy techniques in control and decision making for complex systems* was scheduled for Thursday 16th at 10:45am. On Thursday the day dawned cooler than previous days as a sign of the fresh breeze of research we got that day. The talk was very brilliant and quite different from usual plenary talks because the point it was not only to show his achievements but rather than was focused on the problems he faced in nuclear complex systems and how they can be solved by using fuzzy techniques both in control and decision making. After the conference ending he was one more day with us in Jaén, and we could discuss about future collaborations and share with him nice moments such as when he bought a typical Spanish music instrument for flamenco dancing, so-called *castanets*, for Greet or different toys for Nicola and Alexander. Eventually I discovered how much he liked Spanish food, mainly seafood, and also many similarities between Spain and China.

Since September 2004 Da and myself was in touch quite often, and in December I received by first time a Da's Christmas greeting that was a photograph from the *Ruan brothers* (Nicolas and Alexander) pretending to play a toy guitar. Thereafter every Christmas I received similar greetings.

He came back to Andalucía on 9th February 2005 (for a seminar held in Granada for Ph.D. students), after his seminar we kept working for a while, in our research, that across the time will produce nine shared journal papers and six conference contributions and other collaborations that have been remarkably important for my research group. On 9th March 2009 Da visited again the University of Jaén as a member of Macarena Espinilla's PhD panel, one of my PhD students. He supported her a lot providing nice compliments and comments to her research and that evening we planned two research stays one of Macarena and another one of Rosa Mª Rodríguez another PhD student. That day Da also met my father and they got along very well despite communication problems because of my father, he only speaks Spanish. However, they could understand each other quite well, he was always very kind with all of us and with our people.

Da was in Jaén on 8th May 2011. That Sunday Da came from Madrid just to visit me and get to know the new ICT Research Institute in the University of Jaén, I became its director among other issues thanks to my international links in which Da played a key role. We enjoyed the day together and of course we made plans for the future in order to keep collaborating and make use of the institute for shared research issues. I keep him in mind as an active researcher whenever and wherever.

Unfortunately that was his last time in Jaén, my feeling is that I have lost one of my best friends and better colleagues and, writing this chapter I could realize how important Da was for my development as researcher and how much support he provided me and my research group during these years.

One of my regrets is that I have no photographs with him in Jaén, so I attached below one with some of my colleagues of Jaén during FLINS 2008 in Madrid.

F Mata, P.J. Sánchez, D. Ruan and L. Martínez in FLINS2008

The previous picture does not include Rosa and Macarena because they asked me to be part of this chapter reporting their memories and expressing their thanks to Da because he was also very important for their careers though they are still very young. In the coming pages they will describe their memories with him.

Eventually, just say that I will miss Da forever. However I will keep him in my heart because he was a brilliant researcher and an excellent friend.

Thanks a million, my friend

Macarena's Memories

Even though my research career is not that long, Prof. Da Ruan has been present throughout it from the beginning, being a crucial person. In 2007, one of my early publications was in a book edited by him and others, the following one was in September 2008, when I attended the FLINS2008 conference held in Madrid. My first impression about Prof. Da was a man full of energy and always smiling.

Undoubtedly, the most intense year with Prof. Da was 2009. First, he was part of my PhD panel in March 2009. I will never forget the moment after the public defense of my PhD, my father and mother were very moved, Prof. Da came to my parents and he congratulated them with some words in Spanish! This was a significant moment for all of us. Prof. Da stayed with us a few days in Jaén and he invited Rosa and me to collaborate with him in the SCK•CEN, Mol, Belgium. To do so, we would work in the center for a month. It was a great opportunity although initially it was a bit difficult for me to think that I would be away from home for a month. In August 2009, I was working in the SCK•CEN and it was wonderful, both professionally and personally. I could collaborate with a magnificent researcher and meet an excellent person, husband and father. Prof. Da gave me support any all time, for example he provided me an accommodation in a residence nearby the center in which young researchers lived. There, I became good friend of other researchers from different countries with whom I shared some experiences. This fact helped me to open my mind and aroused my desire to know other places, cultures, etc.

Despite the distance, we kept collaborating; Prof. Da was always very supportive with us. In September 2010, I attended a conference about e-learning again in Hasselt, Belgium. Prof. Da came from Mol to Hasselt to meet me. We were having lunch together and we discussed about our research. That was the last time I was with Prof. Da, it was very nice. I keep in my mind with love that last moment and fortunately I have a photograph of that occasion.

D. Ruan and M. Espinilla in Hasselt, 16th Sept. 2010

I want to express my thanks to Prof. Da in this chapter, I will be eternally grateful to Prof. Da for contributing to my professional and personal development.

Thank you so much for everything Prof. Da

Rosa's Memories

I met Prof. Da, first time, on 9th March 2009 when he came to the University of Jaén as member of the PhD panel of my colleague, Macarena. I remember that I had brought some cakes boxes from my village to give them to the members of the PhD panel. Therefore, when my supervisor, Luis Martínez, introduced me to Prof. Da, I gave him a cakes box and I explained him, what they were made of and how many types there were. I was very nervous because I had to speak in English and I knew that Prof Da was an important researcher. He was very kind with me and he made an easy conversation, I then realized that Prof. Da was not only a brilliant researcher but also an excellent person. That day was a remarkable day for Macarena, but I never thought that it would be also for me. In the evening I met Prof. Da again because the research group went to have dinner altogether. Before the dinner, Prof. Da proposed me going to Mol for one month to develop a software application and carry out some experiments about nuclear safeguards. It was a great challenge for my career, so I accepted very grateful.

The stay was very nice, because Prof. Da helped me to feel comfortable working with him and he was very kind all the time. Besides, the results obtained were successful and have produced three shared journal papers and two conference contributions which have been very important for my career and for our research group.

D. Ruan and RM. Rodríguez in Jaen University in 2009

On 22nd August 2009, I went to the University of Ulster in Belfast for attending to an Advanced Course in Artificial Intelligence, and Prof. Da was one of the speakers. His talk about *compensate or not compensate fuzzy aggregations in complex decision systems* was divided into four days, therefore, he had time to have a cup of tea with me and talk about future collaborations. I attached below a photograph with him in the Parliament Building Stormont in the gala dinner of the course.

Unfortunately, I will not be able to listen to him in another talk and I notice how important Prof. Da has been for my career. For this reason, I want to express my thankfulness to Prof. Da in this chapter, because he supported me a lot since I started my PhD in 2009.

Thank you very much Prof. Da

3.10 Professor Da Ruan's Academic Activities with His Turkish Friends

Prof. Dr. Cengiz Kahraman and Dr. Özgür Kabak

Istanbul Technical University, Department of Industrial Engineering, İstanbul, Turkey

Prof. Cengiz Kahraman first contacted Prof. Da Ruan during the 5th International FLINS Conference, which was held in Gent, Belgium on September 16-18, 2002. Prof. Kahraman tells this contact period as follows: "*When I informed Prof. Da Ruan about the problem, which prevented my travel to Belgium for FLINS 2002 conference, I was expecting an angry response from him. On the contrary, he understood my problem and said "It is no problem. We would like to see you at the next FLINS conference. I will mail the FLINS 2002 conference proceedings and materials to your postal address.*" This was the first contact between Prof. Kahraman and Prof. Da Ruan.

Prof. Da Ruan with his Turkish friends in Toledo in 2008

The next FLINS conference was held in Blankenberge, Belgium on September 1-3, 2004. Prof. Kahraman attended at this conference together with his colleagues Ozgur Kabak, Tijen Ertay, Gulcin Buyukozkan, and Ufuk Cebeci. This was the beginning of a sincere and warm relation between Prof. Da Ruan and his Turkish friends. Prof. Da Ruan welcomed each of the Turkish participants individually and introduced the organizers of FLINS 2006. They were Martine De Cock and Mike

Nachtegael and the others. Prof. Jie Lu was the person who Prof. Da Ruan introduced to Turkish friends as one of his best friends.

The next FLINS conference was held in Genoa, Italy on 29-31 August 2006. Prof. Kahraman participated at this conference with his colleagues Ziya Ulukan, Tijen Ertay, Alp Ustundag, Nufer Yasin Ates, Ahmet Beskese, Esra Albayrak, Yasemin Erensal, Sezi Cevik, Gulcin Yucel, Emel Aktas, Mehmet Tanyas, İhsan Kaya, Didem Cinar, Gulcin Buyukozkan, Orhan Feyzioglu, Sezgin Kilic, Tunc Bozbura, Tufan Demirel, and Nihan Cetin Demirel. Prof. Da Ruan awarded Prof. Kahraman FLINS Outstanding Service Award.

The next FLINS conference was held in Madrid, Spain on 21-24 September 2008. The Turkish participants were Cengiz Kahraman, Alp Ustundag, Emre Cevikcan, Serdar Kilinc, Hakan Tozan, Ozalp Vayvay, Ozgur Kabak, Ihsan Kaya, Gulcin Buyukozkan, Cagri Tolga, Mujde Genevois, Ziya Ulukan, Ufuk Cebeci, Adem Karahoca, Selcuk Cebi, Nihal Erginel, Sevil Senturk, Esra Bas, Irem Ucal, Basar Oztaysi, Tufan Demirel, Nihan Cetin Demirel, Orhan Engin, and Hulya Behret.

The 9th FLINS Conference was held in Chengdu, China on 2-4 August 2010. The Turkish participants were Cengiz Kahraman, İhsan Kaya, Gulcin Buyukozkan, Gizem Cifci, Orhan Feyzioglu, Irem Ucal, Can Ucuncuoglu, Ziya Ulukan, Umut Asan, Ayberk Soyer, Cagri Tolga, Pinar Dursun, Tolga Kaya, Hulya Behret, Nihan Cetin Demirel, Tufan Demirel, Ahmet Can Kutlu, Serdar Kilinc, Seda Yanik, Tijen Ertay, Seyda Serdar Asan, Tuncay Gurbuz, and Alp Ustundag. The FLINS participants' number from Turkey significantly increased from 5 (FLINS 2004) to 24 (FLINS 2010) in six years.

Prof. Da Ruan was always open to scientific collaborations and ready to perform the most of the collaboration work. He never tried to avoid burdensome works and did his best in collaborations. He always became a leading man for his Turkish friends by bounteously letting them utilize his experiences. Prof. Kahraman tells that Prof. Da Ruan was the man who taught him how to reveal someone's own talents and encourage to exhibit them in international arena. Prof. Kahraman also indicates that Prof. Da Ruan always reviewed his Turkish friends' publications and proposed corrections before they went into the submission process, even he was not a co-author in these works. Prof. Da Ruan was absolutely selfless, sincere, and patient. Prof. Kahraman believes that nobody could live an experience, which he/she could receive no response to an e-mail message sent to Prof. Da Ruan. He always replied them even if he was in travel or at a conference or on vacation.

Prof. Da Ruan worked with Prof. Kahraman in the following topics: quality function deployment (Ertay et al., 2007; Buyukozkan et al., 2004a), cash flow analysis (Kahraman et al., 2000; Kahraman et al., 2002; Kahraman et al., 2003a; Kahraman et al., 2004c, Beskese et al., 2003), control charts (Gülbay et al., 2004), multi-criteria decision making (Buyukozkan et al., 2004b; Kahraman et al., 2004a; Kahraman et al., 2003b; Bozdag et al., 2003; Kahraman et al., 2004e), software development (Kahraman et al., 2004d), and statistical analysis (Kahraman et al., 2004b)

Prof. Kahraman's latest collaboration with Prof. Da Ruan, which was almost completed, was the co-guest-editorship of the special issue of International Journal of Applied Management Science (IJAMS). This special issue was on "Modern Tools of Industrial Engineering: Applications in Decision Sciences". Prof. Kahraman wrote the final paragraph of this issue's preface as follows: "*While accepting this special issue project, we did not know what the future would bring to us. As our project is almost completed, I lost my dear friend and my co-guest editor Professor Da Ruan on July 31, 2011. He was only 50 years old. He was an excellent researcher and wonderful, kind and helpful person. His friends and I will always miss him.*"

Prof. Ruan with Dr. Kabak in Mol in 2010

Dr. Özgür Kabak first met Prof. Da Ruan during the FLINS 2004 in Blankenberge, Belgium. Five years later, he held a postdoc fellowship from Belgian Science Policy Office (BELSPO) with Prof. Da Ruan as the promoter for a nuclear safeguards evaluation project. Dr. Kabak studied on the project under the supervision of Prof. Da Ruan at Belgian Nuclear Research Centre (SCK·CEN) for one year starting from February 2009. During his stay in SCK·CEN he had a great chance of familiarizing with Prof. Da Ruan in both scientific and personal manner.

The aim of the project entitled "A fuzzy multiple attribute decision-making approach (FMADM) for nuclear safeguards information management" was to propose a FMADM approach for handling and treatment of nuclear safeguards related information. During the one year research period this aim and more have been achieved.

The project introduces a new approach depending on cumulative belief degrees (CBDs) for dealing with different types of scarce data which is one of the most important problems in nuclear safeguards evaluation (NSE). Nuclear safeguards are a set of activities by which International Atomic Energy Agency (IAEA) seeks to verify that a State is living up to its international undertakings not to use nuclear programs for nuclear weapons purposes. In NSE, indicators are evaluated by the

IAEA experts on the basis of their analysis of the available information sourced from declaration of States, on-site inspections, IAEA databases of non-safeguards, and open sources such as Internet and newspapers. The evaluations are made in different time periods and by different experts. Therefore expert evaluations of the indicators for different periods are to be aggregated to make a final decision. Since this process contains subjective judgments of the experts and aggregation of multiple evaluations, it is complicated under various uncertainties.

Prof. Ruan with Dr. Kabak's son, Arda, in Mol in 2009

The proposed CBDs methodology (Kabak and Ruan, 2011a) aims to aggregate available expert judgments about the indicators for different aspiration levels. The authors used the available data and did not delete or impute the missing values. Additionally, a measure intended to give the idea of amount of data in which the result is based on. For these purposes, the model uses fuzzy linguistic terms for the expert evaluations of indicators. The belief structure, which enables getting the evaluations with distributions over the linguistic terms, is suggested to represent the expert evaluations. Moreover, based on the belief structure, CBD is proposed for aggregating the evaluations at different linguistic term levels. Finally, a reliability index is defined for measuring the extent of available data. The applicability of the approach is illustrated by a numerical example.

Another important output of the project was a FMADM method for the NSE problem (Kabak and Ruan, 2011b). To propose this method, initially, the NSE problem is defined as a FMADM problem. Subsequently the applicability of four basic methods to the NSE problem are investigated: a simple additive weighting (SAW) method, a technique for order preference by similarity to ideal solution (TOPSIS) approach, a linguistic term-based approach, and an operator-based method. Finally a FMADM method is proposed for the NSE problem based on CBDs and order weight averaging (OWA) operators.

Another effort of the project was to develop a software package for applying the proposed CBDs approach. Rolando Quinones from Antonio Narino University, Colombia joined the team for developing the software, thereafter called *AdUnIT*. It is aimed to apply the CBDs approach to different problems by the help of *AdUnIT*. Then energy policy evaluation problem is solved using the software (Ruan et al., 2011).

Dr. Özgür Kabak tells his impressions about Prof. Ruan's scientific personality as follows: "*I feel myself privileged as I had a chance of working with him. I learned a lot from him during my Postdoc research in SCK·CEN. I can describe his scientific personality with these words: hard working, open-minded, helpful, and ever-searching. He was in charge of many projects including nuclear safeguards, energy policy, FLINS, ISKE, IJCIS etc. and he was properly achieving these projects with his accurate planning and hard working. I had never heard any complaints from him about his work. He was an open-minded scientist with whom you can discuss any topic and allows you to convince him with ground statements. He was open to new ideas that enabled us to develop CBDs approach for nuclear safeguards. He had a purely helpful personality that also shaped his scientific personality. He never refuses any scientific help. He always answered my unplanned questions regarding the project respectfully. He was continuously searching for new projects, new research areas and new applications of the developed methods. For instance he directed CBDs approach, which is originally proposed for nuclear safeguards problem, to apply in the energy policy problem. He was also searching for European projects for applying his theoretically developed decision making methods to support European policy makers.*

He was the epitome of hospitality. He was always the good scholar and a great friend. I will miss him and I extend my sincere condolences to his wife Greet and his sons. May he rest in peace."

Prof. Ruan also worked with his other Turkish friends Gulcin Buyukozkan, Tijen Ertay, and Ufuk Cebeci and published many papers (Buyukozkan and Ruan, 2008; Buyukozkan and Ruan, 2010; Buyukozkan et al., 2012; Buyukozkan et al., 2007; Ertay and Ruan, 2005; Ertay et al., 2006; Cebeci and Ruan, 2007).

References

Beşkese, A., Kahraman, C., Ruan, D.: Flexibility quantification in computer integrated manufacturing systems based on fuzzy cash flow analysis. In: Proceedings of 5th International FLINS Conference on Computational Intelligent Systems for Applied Research, Het Pand, Gent, Belgium, September 16-18, pp. 171–178 (2002)

Bozdag, C.E., Kahraman, C., Ruan, D.: Fuzzy group decision making for selection among computer integrated manufacturing systems. Comput. Ind. 51, 13–29 (2003)

Buyukozkan, G., Ertay, T., Kahraman, C., Ruan, D.: Determining the importance weights for the design requirements in house of quality using fuzzy analytic network approach. Int. J. Intell. Syst. 19, 443–461 (2004a)

Buyukozkan, G., Kahraman, C., Ruan, D.: A fuzzy multi-criteria decision approach for software development strategy selection. Int. J. Gen. Sys. 33, 259–280 (2004b)

Buyukozkan, G., Feyzioglu, O., Ruan, D.: Fuzzy group decision-making to multiple preference formats in quality function deployment. Comput. Ind. 58, 392–402 (2007)

Buyukozkan, G., Ruan, D.: Evaluation of software development projects using a fuzzy multi-criteria decision approach. Math. Comput. Simul. 77, 464–475 (2008)

Buyukozkan, G., Ruan, D.: Choquet integral based aggregation approach to software development risk assessment. Inf. Sci. 180, 441–451 (2010)

Buyukozkan, G., Arsenyan, J., Ruan, D.: Logistics tool selection with two-phase fuzzy multi criteria decision making: A case study for personal digital assistant selection. Expert. Sys. Appl. 39, 142–153 (2012)

Cebeci, U., Ruan, D.: A multi-attribute comparison of Turkish quality consultants by fuzzy AHP. Int. J. Inf. Technol. Decis. Mak. 6, 191–207 (2007)

Ertay, T., Buyukozkan, G., Kahraman, C., Ruan, D.: Quality function deployment implementation based on analytic network process with linguistic data: An application in automotive industry. J. Fuzzy. Intell. Sys. 16, 221–232 (2005)

Ertay, T., Ruan, D.: Data envelopment analysis based decision model for optimal operator allocation in CMS. Eur. J. Oper. Res. 164, 800–810 (2005)

Ertay, T., Ruan, D., Tuzkaya, U.R.: Integrating data envelopment analysis and analytic hierarchy for the facility layout design in manufacturing systems. Inf. Sci. 176, 237–262 (2006)

Gülbay, M., Kahraman, C., Ruan, D.: Alpha-cut fuzzy control charts for linguistic data. Int. J. Intell. Syst. 19, 1173–1195 (2004)

Kabak, Ö., Ruan, D.: A cumulative belief degree-based approach for missing values in nuclear safeguards evaluation. IEEE Trans. Knowl. Data Eng. 23, 1441–1454 (2011a)

Kabak, Ö., Ruan, D.: A comparison study of fuzzy MADM methods in nuclear safeguards evaluation. J. Glob. Optim. 51, 209–226 (2011b)

Kahraman, C., Ruan, D., Tolga, E.: Discounted possibilistic cash flows versus probabilistic cash flows. In: Proceedings of The Fourth International FLINS Conference on Intelligent Techniques and Soft Computing in Nuclear Science and Engineering, Bruges, Belgium, August 28-30, pp. 214–223 (2000)

Kahraman, C., Ruan, D., Tolga, E.: Capital budgeting techniques using discounted fuzzy versus probabilistic cash flows. Inf. Sci. 142, 57–76 (2002)

Kahraman, C., Ruan, D., Bozdag, C.E.: Optimisation of multilevel investments using dynamic programming based on fuzzy cash flows. Fuzzy Optim. Decis. Mak. J. 2, 101–122 (2003a)

Kahraman, C., Ruan, D., Doğan, I.: Fuzzy group decision making for facility location selection. Inf. Sci. 157, 135–153 (2003b)

Kahraman, C., Cebeci, U., Ruan, D.: Multi-attribute comparison of catering service companies using fuzzy AHP: The case of Turkey. Int. J. Prod. Econ. 87, 171–184 (2004a)

Kahraman, C., Bozdag, C.E., Ruan, D., Ozok, A.F.: Fuzzy approaches to statistical parametric and nonparametric tests. Int. J. Intell. Syst. 19, 1069–1087 (2004b)

Kahraman, C., Beskese, A., Ruan, D.: Measuring flexibility of computer integrated manufacturing systems using fuzzy cash flow analysis. Inf. Sci. 168, 77–94 (2004c)

Kahraman, C., Buyukozkan, G., Ruan, D.: Measuring software development value using fuzzy logic. In: Ruan, D., Zeng, X. (eds.) Intelligent Sensory Evaluation: Methodologies and Applications, pp. 285–308. Springer (2004d)

Kahraman, C., Buyukozkan, G., Ruan, D.: A fuzzy heuristic multi-attribute conjunctive approach for erp software selection. In: Proceedings of 6th International FLINS Conference on Applied Computational Intelligence, Duinse Polders, Blankenberge, Belgium, September 1-3, pp. 519–524 (2004e)

Ruan, D., Kabak, Ö., Quinones, R.: An ordered weighted averaging operator-based cumulative belief degree approach for energy policy evaluation. Int. J. of Adv. Oper. Manag (2011) (accepted for publication)

3.11 Da Ruan and Research Collaboration between Manchester University and SCK

Prof. Dr. Jian-Bo Yang

Manchester Business School, The University of Manchester, UK

The research link between Manchester and SCK was initially established in 2002 after Dr Jun Liu joined Manchester Business School (MBS) of The University of Manchester (UoM) to work as a Research Associate for a project funded by the UK Engineering and Physical Science Research Council. Having worked with Professor Da Ruan at SCK for a couple of years just before joining MBS, Jun was familiar with Da's research and suggested that there should be common research interests between MBS and SCK. This led to my initial direct contact with Da.

Following many email and telephone communications, Da invited me to pay a short visit to SCK in Mol in 2003. That was my first visit to SCK and I was very much impressed by Da's hospitality and careful planning. He arranged a taxi to collect me from Brussels Airport to SCK. Although I arrived at the SCK guest house rather late in the evening, Da still waited there and showed me how to get into the guest house through the side door and find the room, as the reception had already been closed. Although that was our first face-to-face encounter, we immediately felt like old friends. I had a very pleasant and productive day with Da the next day, providing a seminar for his group in SCK and discussing potential collaborations between MBS and SCK. Da and his colleagues were especially interested in the Intelligent Decision System (IDS) we had developed in Manchester. At the time, Da and his colleagues were involved in a large project for which it was challenging to estimate costs for constructing an experimental nuclear reactor due to many uncertain factors that must be taken into account for the cost estimation. Da suggested whether we could tailor-make IDS to help fulfill the task, which set up a target for our initial research collaboration. A new version of IDS, referred to as IDSCostimator, was later developed and used by SCK for supporting the cost estimation. During the process, Da and his colleague visited Manchester and I visited SCK again for the delivery and training for the use of the IDSCostimator software. The photo below recorded my reunion with Da in Mol in 2005.

The research between MBS and SCK has since been widened to many areas, in particular nuclear energy policy analysis, safety culture assessment, nuclear safeguards, group decision support systems, etc. In April 2010, Professor Dong-Ling Xu and I paid a short visit to Da in Mol for planning our future research collaboration between MBS and SCK. We drove from Manchester to Mol on route to attend a conference in France. Due to the unexpected delay to cross the English Channel by ferry, we arrived at SCK near midnight. Da was still waiting for us at the SCK guest house, smiling as usual and talking with us so that we didn't feel tired at all after the long drive. We spent the next day to discuss our future plan of collaboration and decided that we should spend more time in SCK to discuss in great detail the challenging real issues faced by SCK. This short visit indeed led to our longer visit to SCK for about one month in August-September 2010 for developing joint research projects and planning joint proposals for funding. We had intensive discussions in many areas and were all excited about the perspectives and potential of collaboration between MBS and SCK, looking forward to long term joint research in next 10 years and beyond. We took immediate actions to look for funding separately from MBS and SCK with success. Da secured funding for a research officer and a PhD student from SCK and we found grants from MBS/UoM for a number of PhD students, who were expected to start collaborating from September 2011. Da agreed to visit Manchester again in late 2011 for further discussing our joint research and providing a seminar to our research centre. While visiting Complutense University in Madrid, Spain in April 2011, Da initialized a meeting for developing joint EU proposals in line with our discussions in Mol in 2010. Dong-Ling and I went to the meeting. That was our last face-to-fact meeting with Da, although we kept in contact with each other via email even when we were attending the IFORS conference in Australia in July 2011.

Da, Jianbo and other friends in UK

It was a terrible shock to hear about Da's sudden departure just a few days after our return to Manchester in early August 2011 following our visits in Australia and China. No word could express our shock at the sad news. Looking forward, however, we will continue our research as discussed with Da and hopefully generate interesting and useful results, which we believe is the best way to remember Da.

3.12 Memories in University of Ulster

Dr. Jun Liu[1], Prof. Dr. Liam Maguire[2], Prof. Dr. Hui Wang[1], and Dr. Juan Augusto[1]

[1] School of Computing and Mathematics, University of Ulster, Northern Ireland, UK
[2] School of Computing and Intelligent Systems, University of Ulster, Northern Ireland, UK

Jun's Memories–Da Ruan: My Mentor, My Friend, and My Inspiration

Words and tears could not express how shocked I was to learn the sad news of Da's passing when we just met each other in April in Madrid for a network meeting and still had some telephone conversations and email contacts afterwards, and looked forward to meeting again for the forthcoming conferences and working together on the EU projects proposals. A few months passed, for me, Da seems still staying with us in my heart, even for now, sometimes, I was still thinking of calling him and sending an email to him. When suddenly realized of his passing, deep pain and losing feeling swamped me … It is really hard for me to condense my memories of him into just a few words. I lost a great mentor and a genuine and dependable friend. I will miss you forever, Da!

I met Da, for the first time, twelve years ago when I was visiting Gent University with my PhD supervisor Prof. Yang Xu hosted by Prof. Kerre. It was an international conference held in Gent in 1999 where I was introduced to Da who become later on my mentor for my Postdoc research in SCK from February of 2000 and influenced my career and life beyond any expectations since then.

Jun and Da in September 2000 at Dresden in Germany

With the passing of time I strongly felt his passion to work and his dedication to his fellows. I found him to be someone who is always easy going, gentle, warm and open hearted and friendly person filled with kindness and full of energy and humour always making the friends around him laughing. He is also an excellent, passionate, dedicated scientist with profound knowledge, lots of patience and persistence. We should feel lucky, because of Da, many researchers and scientists including all of us sit together and then become good friends within a wonderful research network leading by him. Da is also a tireless and serious worker. I was always greatly impressed with his professionalism and seriousness in work. He replied every single e-mail no delay, read and corrected each single word of the papers or thesis upon been asked. I remembered so many times that I received his emails and polished papers after midnight.

I was privileged to have Da as a mentor. Da was a great mentor guiding me in my research while I worked in SCK at Mol, and continuously supporting and inspiring me all the ways in developing my research and career path (from Mol to Manchester to Belfast). Da was an inspiration who was there whenever a clever advice was needed, always encouraging me to look at the wider global picture from theoretical to practical aspects, as he put it. I still remember clearly the first day of my working in SCK while we met in his office 11 years ago. He kindly spent over an hour introducing his research activities and inspiring me to be a good researcher. He particularly and continuously inspired and supported me to get involved in research development, research community and network. All these have been carved deeply into my memory, experience and affected my thinking of doing research throughout the way.

Since I left Mol, we have kept continuously working together for different research papers, books and project proposals. He has visited Manchester and Belfast for setting up research link and giving the invited talk and lectures. He was always there to share experience and give support. He actually treated all his fellows and friends with care and supported us as much as he could. I will never forget our last conversation in his rented apartment in Madrid in April after the project network meeting. I was really touched of the conversation and follow up emails and telephone conversation offering different funding opportunities, support and kind advice in terms of working and life, and experience shared with me. For years, I realized how important in my career life.

Da was a great friend caring and supporting me and my family a lot during my stay in Mol in every aspect, especially before and after my son was born. He has been like a specially hired driver brought us to and from the hospital for regular check until my son was born and I got my driving license. To me, Da and his family has been a lifelong friend of my family. We still remember clearly when my wife and I met Da in Brussels airport. We arrived very late about 11pm. It takes about 2 hours driving from Brussels to Mol. In the following days, he had been helping us sorted out all the issues as a new arrival to Belgium. Since then, we started to get to know Da and his family. We could recall the times both our family together in various events (starting with the first welcome dinner in SCK club house), we had lots of interesting chats together, also had lots of fun together making dumpling, as we all are food lovers. In various festivals (Middle Autumn Festival and Spring Festival), we enjoyed the talking and eating while the kids played together and had lots of fun.

Da, I just wanted to thank you for everything, as a great mentor, as a great friend and as an inspiration. You will always live within me and I will always be grateful. Here on behalf of my family, I want to express our deep appreciation and respect of you for your kindness, willingness and sincerity in helping and supporting me and my family all the ways since we met.

Just one more thing, Da has always talked about how proud he is of his sons, about many interesting things that they are doing. Although our son met Da and his family only a few times since we left Mol, I hope that our next generation will make more connections and be good friends, to continue a legacy in memory of Da.

Da, I will never forget the experiences we've been together, I will never forget your great contributions to our research community, your great support and help, your energy and enthusiasm, your working spirit and kindness, and smile! I will miss you forever!

Prof Da Ruan, Prof Yang Xu, Prof. Keyung Qin and Dr Jun Liu in Ghent at FLINS2002

Liam's Memories

I first met Da in 1995 at the IFSA World Congress in Sao Paula, Brazil. This was my first attendance at this series of conferences and I met him at breakfast in the hotel one morning. He then introduced me to a number of his associates as he had organized the conference a few years earlier in Belgium. I will always remember that initial breakfast and the warm welcome that he extended to me. We remained in contact since then, mostly via emails, and we also met up at a number of conferences since then. He also invited me to contribute a number of book chapters for his edited volumes and he invited me onto a number of programme committees. He has always been very interested and supportive of my own research. Da was a great researcher and more importantly he was an excellent human being. I wish I had more opportunities to enjoy his company. He will be sorely missed. The following Irish saying/poem is very appropriate:

Tá daoine a shiúlann inár saolta agus shiúlann amach astu go luath
Tá daoine a fhanann ar feadh tamaill
Agus fágann siad rianta a gcos ar ár gcroíthe
Agus casann ár n-anamacha port nua go deo deo.

And translated in English as

Some people come into our lives and quickly go
Some people stay awhile
And leave footprints on our hearts
And we are never, ever the same

Hui's Memories–In Memory of Da Ruan: a Scholar, a Visionary and a Friend

It's more than 3 months since Da's sudden passing away. Every time I think of it, my heart is still hurting and I am still feeling sorry for the early end of a journey with a bright future. At the age of 50, he was still young and energetic, and he must still have lots of dreams unfulfilled. What a pity!

I knew his name a long time ago, but I didn't know him personally for too long. I met him three times in recent years, twice at the University of Ulster and once in Brussels. The first time I met him was in March 2007 when he visited the University of Ulster. He gave a talk entitled "Applied AI in Complex System" (1:15pm, 20 March 2007, Room 16D05) in the School of Computing and Mathematics. He visited the University of Ulster the second time in July 2007. We discussed collaborations in joint EU proposals relating to policy decision making for sustainable environments and also contributions to a book titled "Intelligent Decision and Policy Making Support Systems" (Springer, 2007). He had been the driving force behind the research proposal initiative, which was gaining momentum shortly before that day. The book was already published but the very person behind it is not here any longer.

He returned to the University of Ulster in 2009 as a keynote speaker at the European Summer School in Artificial Intelligence (Advanced Course in Artificial Intelligence -- Intelligent Decision Support Systems, 23-29 August, 2009) of which I was a co-chair (together with Dr Juan Carlos Augusto), and gave a series of lectures titled "To compensate or not to compensate: fuzzy aggregations in complex decision systems". That was the longest period of time (4 days) that I spent with him.

Shortly after the Summer School, Dr. Jun Liu and I went to Brussels to attend the Security Mission Industry Group (SMIG) meeting (14 & 15 September 2009) for participants in the European Security Research Programme. This was the home country of Da, and he showed his hospitality by taking me and Jun to lunch at a very good restaurant close to ASD (AeroSpace & Defence Industries Association of Europe), the organiser of the SMIG meeting. I cannot remember the name of the restaurant but I will definitely visit it again next time I go to Brussels. At this meeting, we together met Mr Dominic Kelly from CBRNE Ltd, UK, who had subsequently invited Da to join the Industrial Mission Group (Security CBRNE Technical Area).

I write here to provide an account of the few days I shared with Da, to help provide a fuller picture of his life and to join the many colleagues and friends in celebrating his life and achievements.

If he could hear me, I would say a few words. Da, you departed too early. However I feel you had lived your life to the full – you left behind your thoughts, your visions, your love and your friendships. You had made a difference! You should be proud of your achievements and the fact that so many people felt empty without you and so many people remember you. May you rest in peace, under the brushy Bodhi tree in Heaven.

Dr. Juan Augusto has added the following paragraph in relation to ACAI'09: Da was one of the invited lecturers for the 2009 edition of the prestigious 'Advanced Course in Artificial Intelligence' which centered on 'Intelligent Decision Support Systems [theory, algorithms, and applications]' and took place at the University of Ulster during August 23-29 of 2009. His course 'To compensate or not to compensate: fuzzy aggregations in complex decision systems' was one of the best attended and most appreciated where students were able to learn about solutions to real life decision-making problems firmly supported by rigorous scientific concepts, complex content that only a scientist of the stature of Da can deliver with confidence.

Hui and Da in March 2007 at University of Ulster, Norther Ireland, UK

Da lecturing at ACAI'09 Summer School on August 23, 2009 at University of Ulster, Northern Ireland, UK

3.13 Da and His Friends in France

Prof. Dr. Xianyi Zeng[1], Prof. Dr. Ludovic Koehl[1], Prof. Dr. Mauricio Camargo[2], and Prof. Dr. Bernadette Bouchon-Meunier[3]

[1] The ENSAIT Textile Institute, Roubaix, France,
[2] ERPI, Institut National Polytechnique de Lorraine, Nancy, France
[3] Laboratoire d'Informatique de Paris 6, University Pierre et Marie Curie, Paris, France

The first week of August 2011 was certainly destined to be full of sorrow for the scientific research community of artificial intelligence. A message "Our best friend Prof. Da Ruan suddenly passed away ..." was repeated many times on our computer screens. Nobody could believe his/her eyes when reading these words. And it is very difficult to describe our feeling since these days: shock, sadness, a sense of fear... Even with the actions of condolence and commemoration, our heart could not become more relaxed. As Da's friends and research partners in France, we often recall the moments when we stayed together with Da in different occasions: international conferences, research meetings, or visits of friendship...

The cooperation between ENSAIT and Da and his research team in SCK-CEN started around 1995. Xianyi Zeng clearly remembers their first meeting during FUZZ-IEEE 1995 in Yokohama, Japan. Da's original view on research activities and his passion for fuzzy theory and its applications has touched many participants. Partly encouraged by his success stories in applied research projects, the researchers of ENSAIT have been working on modelling of human factors with fuzzy logic and other intelligent techniques for product design optimization and market exploitation.

After the conference in Yokohama, the researchers of ENSAIT consecutively received Da's invitations for participation in FLINS conferences. This is a series of international conferences on computational intelligence, initiated and organized by Da. We have participated in almost all the FLINS conferences since 1998. Through these conferences, we got to know a great number of theoretical and applied scientists working in similar research areas. Many significant discussions on common research interests were conducted by Da in both formal and informal meetings. During these meetings, Da fully showed his personal characters, i.e. dynamism, quick reaction, flexibility, sensitivity, opening and communication spirit. He was so sensitive to each discussion topic and capable of conducting the meeting in the right orientation.

Da's research interests covered many scientific areas, including modelling of textile and clothing products, rather far from his original research competences. During a long professional career in scientific research, he was formed to be quickly adapted to any application contexts and find relevant solutions with his specialized competences. In 2002, he gave a lecture at ENSAIT on computational intelligence in industrial applications. His lively speech with real examples impressed many colleagues and doctorate students. In 2003, he was invited to give a plenary presentation in SEIT03, jointly organized by ENSAIT and Donghua University in Shanghai. It was an international workshop on sensory evaluation using intelligent techniques. Since then, he collaborated with us on sensory analysis of industrial products and obtained a number of results. Especially, in collaboration with the researchers of ENSAIT, Da co-edited two books, entitled "Intelligent sensory evaluation: theory and applications" and "Computational Textile", both published by Springer in 2004 and 2007 respectively.

Zhong, Jie, Da and Xianyi in a research meeting in SCK, Aug. 2007

In 2008 and 2009, as professor at Ghent University, Da was involved in a research project of GRAISyHM (a research group gathering all researchers in automation in the French Region Nord-Pas de Calais). Our project, entitled "interpretation and decision of complex industrial systems by integrating numerical and linguistic data", aims at optimizing products in order to meet consumer's perception and functional requirements. The main partners in this project include Ghent University (Da Ruan and Etienne Kerre), ENSAIT (Ludovic Koehl and Xianyi Zeng), Mauricio Camargo and Christian Fonteix (ERPI of Nancy). In the frame of this project, Da took part in all the research meetings and participated in the supervision of one doctorate student working on modelling fashion themes and their relations with textile technical parameters for two months. As result of our collaboration in this project, we published one joint paper in a peer-reviewed journal and presented another one in an international conference.

The cooperation between Da and the Fuzzy Logic team of the Laboratoire d'Informatique de Paris 6 (LIP6) at the Université Pierre et Marie Curie found its roots in their joint participation in European and international conferences on fuzzy logic and intelligent systems during twenty years. The last one was the

World Conference on Soft Computing, held in San Francisco in May 2011, where we could not guess that we were seeing him for the last time. Their relations were reinforced by his participation to several issues of the International Conference on Information Processing and Management of Uncertainty in knowledge-based systems (IPMU) the LIP6 Fuzzy Logic team organized or supported in Paris (1998), Annecy (2002) or Dortmund (2010). In return, several members of the team (Bernadette Bouchon-Meunier, Christophe Marsala) were very glad to take a part in ISKE or FLINS conferences Da organized, being particularly impressed by the activity of the group he was cooperating with in China, led by Tianrui Li, and in Australia with Jie Lu. The scientific community around him was an example of friendly relationship, excellent organization and scientific quality and it has always been a real pleasure to collaborate with them. They are now preparing a special issue of the International Journal of Uncertainty, Fuzziness and Knowledge-based Systems whose editor-in-chief is Bernadette Bouchon-Meunier, dedicated to Computational Intelligence for Policy Making and Risk Governance, which Da had planned to co-edit and which will be introduced by a document prepared by his community in appreciation of his humanity and his scientific achievements.

Tianrui, Da, Bernadette,Guangquan and Yang in Chengdu in 2010

The FLINS 2010 conference, held in Chengdu in August 2010, was an exceptional moment of scientific exchanges and empathy, remembered with emotion by his French friends from the LIP6 and GRAISyHM present there. Also, IFAC HMC 2010 was the last meeting between Da and his French partners of

GRAISyHM. In this conference, we jointly organized a workshop on Human-Centred Industrial Design at the beginning of September 2010 in Valenciennes, France. During three days, in the comfortable climate and quiet atmosphere of Northern France, we enjoyed local foods and talked about many topics: research, industry, fashion design, technology, style of life, Europe, Asia … Each of us can still clearly remember these discussions but Da has gone forever …, leaving us with endless sorrow and regret.

3.14 Da in Donghua University

Prof. Dr. Yongsheng Ding

College of Information Sciences and Technology, Donghua University, Shanghai, China

Getting to Know Da

From March–May 2003, I have been as a visiting professor at Professor Xianyi Zeng's lab, Laboratoire de Génie et Matériaux Textiles, Ecole Nationale Supérieure des Arts et Industries Textiles, Roubaix, France. One day Professor Zeng and I visited Professor Da Ruan's lab at Belgian Nuclear Research Centre (SCK•CEN), Mol, Belgium. We had a good talk together, and became good friends from then. Also, from that talk, I knew that Da had known Professors Shihuang Shao and Jikai Ding at Donghua University before.

Actually, before this talk, I knew that Professor Da Ruan was one of top scientists in the world in the fields of computational intelligence, computational mathematics and nuclear science, as well as the chairman of FLINS (Fuzzy Logic in Nuclear Science) at the Belgian Nuclear Research Centre (SCK•CEN).

Collaboration in Research

At the beginning of 2004, our application for the 5th Shanghai specialized founding for outstanding overseas Chinese scholar was approved successfully by Shanghai Human Resource Bureau, China. In June and July, supported by this sponsorship, Professor Da Ruan was invited to visit the College of Information Sciences and Technology at Donghua University and engaged as an Advisory Professor of the Donghua University.

During this period, he also accepted the invitation from the graduate school of our university, and made a public academic speech entitled "how to conduct scientific research and publish academic papers in international journals" to all graduate students of Donghua University. The speech was successful and aroused everybody's strong interest, so that many students hoped that they could have more opportunities to receive scientific instructions from Professor Da Ruan. To this end, the leaders of Donghua University exchanged ideas with him, expressing the expectation of further cooperation. Professor Da Ruan was delighted that he could contribute more to Chinese universities.

Hereafter, every year he came to our university to give 2-3 academic speeches to teachers and students, hosted several symposia helped us to coach pupils, coauthor academic papers, and so on.

Da was also an excellent advisor who stimulated my students' passions in science and successfully guided them to maximize their potentials in research. Under his guidance, these students learned much about not only how to conduct research but also how to communicate scientific ideas. My students and I deeply appreciate his help as his input has positively affected their professional capabilities.

Da's tremendous intelligence and broad knowledge offered me and my students with the best possible research consultation. His suggestions about practical issues of our research allowed me to think much more deeply about the application of the work. His advice on our work greatly helped us to shape our research work.

Under his help, our collaborated research achievements have been published in International Journals, such as "Information Sciences", "Dynamics of Continuous, Discrete & Impulsive Systems, Series B: Applications & Algorithms".

Prof. Da Ruan as the Advisory Professor of Donghua University

Moreover, Professor Da Ruan helped the Donghua University obtain the opportunities of collaborating with some universities in Spain, Belgium, Germany, Britain, and China for applying for the Sixth European Community Framework Programme and other international projects. With Da's help, the Donghua University, Belgium's Ghent University, Hasselt University, and China's Tsinghua University successfully applied for a Chinese - Belgium bilateral cooperation project entitled "Intelligent systems for data mining and information processing: methods and applications" (No. 011S1105). The success in applying for the above projects has positively influenced the discipline building of the Donghua University. In order to promote the research efforts in the above projects, we decided to host an academic conference in Shanghai - 2006 International Conference on Intelligent

System & Knowledge. This conference was also sponsored by the International (regional) Collaboration Program from National Natural Sciences Foundation of China (No. 60610306301), the Kuancheng Wong Education Foundation of China, the Shanghai Baiyulan technology talents fund from Science and Technology Commission of Shanghai Municipality, China, and the fund from Shanghai International Culture Association, China.

Prof. Da Ruan giving a talk at Donghua University

Academic Conferences

We have held the 2003 Sino-French Workshop on Sensory Evaluation and Intelligent Techniques, Dec. 17~19, 2003, Shanghai, China. Professor Da Ruan was invited to give us a keynote speech "The Role of Computational Intelligence in Sensory Evaluation" at this conference.

With the help of Professor Da Ruan, we have also hosted the 2006 International Conference on Intelligent System & Knowledge (ISKE2006), April 6-7, 2006, Galaxy Hotel, Shanghai, China. The co-organizations included Donghua University, Shanghai, China; Southwest Jiaotong University, Chengdu, China; Tsinghua University, Beijing, China; Ghent University, Gent, Belgium; Belgian Nuclear Research Centre (SCK•CEN), Mol, Belgium; Hasselt University, Hasselt, Belgium; University of Technology, Sydney (UTS), Australia; Southwest University of Finance and Economics, Chengdu. ISKE2006 aimed at covering the state-of-the-art research and development in all aspects related to intelligent systems for knowledge engineering with its missions: 1) Discussing research on intelligent systems for solving problems in knowledge engineering partially under the 2005-2006

China-Flanders Project (No. 011S1105): "Intelligent Systems for Data Mining and Information Processing: Methods and Applications." 2) Bridging the gap between computational intelligence and knowledge systems via joint research between China and international research institutes and universities. 3) Encouraging interdisciplinary research and bringing multi-discipline researchers together on intelligent systems for knowledge engineering and its related fields. Over 200 researchers, professors, and students from 12 countries and regions participate in ISKE2006 and communicated via a total of 16 regular sessions, 3 keynote lectures and 2 featured lectures. The keynote speakers included: 1) Professor Lotfi A. Zadeh from University of California, Berkeley, CA, USA, whose topic was "From search engines to question-answering systems-the problems of world knowledge, relevance, precisiation and deduction"; 2) Professor Ronald R. Yager from Iona College, USA, whose topic was "Some learning methods for data mining and intelligent system"; 3) Professor Da Ruan from The Belgian Nuclear Research Centre (SCK•CEN) & Ghent University, Belgium, whose topic was "Applied artificial intelligence in complex systems"; 4) Professor Jorge Samuel Benitez-Read from National Nuclear Research Institute of Mexico, whose topic was "Fuzzy control and intelligent systems applied to research nuclear reactors and NPP"; and Professor Jie Lu from University of Technology, Sydney, Australia, whose topic was "E-service personalization: concepts, systems, and applications".

With the hard work of Professor Da Ruan, the ISKE conference has become a series of International conferences, and the ISKE'2011 was held this year in Shanghai, China.

Prof. Y. Ding, Prof. L. A. Zadeh and Prof. D. Ruan at ISKE 2006

3.15 Professor Da Ruan in Shanghai University

Dr. Yang Banghua and Dr. Fei Minrui

School of Mechatronics Engineering and Automation, Shanghai University, China

I knew Prof. Da Ruan in 2007. At that time, Prof. Da Ruan was invited as a keynote speaker of 2007 International Conference on Life System Modeling and Simulation (LSMS2007) organized by Shanghai University. The conference was held on September 14-17, 2007, Shanghai, China. I was in charge of contacting keynote speakers as one of program committee members. I had many contacts with Professor Da Ruan by email. Prof. Da Ruan always replied me in time. Sometimes he replied me within one minute after I sent him an email. I felt he was energetic and was always working beside his laptop, which gave me a deep impression. Though he had not attended the LSMS2007 because of a conflict with another conference, he gave us a lecture titled "Lessons learned from computational intelligence applications in complex systems" on September 24, 2007 after the conference. Since then he was kindly engaged as an adjunct professor of shanghai university because of his great achievements, talent and intelligence. The abstract of the lecture "Lessons learned from computational intelligence applications in complex systems" was as follows:

In recent years there has been a growing interest in the need for designing intelligent systems to address complex engineering problems. One of the most challenging issues for the intelligent system is to effectively handle real-world uncertainties that cannot be eliminated. These uncertainties include sensor imprecision, instrumentation and process noise and disturbances, unpredictable environmental factors, to name a few. These uncertainties result in a lack of the full and precise knowledge of the system including its state, dynamics, and interaction with the environment. Computational intelligent techniques including fuzzy logic, neural networks, and genetic algorithms etc., as complimentary to the existing traditional techniques, have shown great potential to solve these demanding, real-world problems that exist in uncertain and unpredictable environments. These technologies have formed the foundation for intelligent systems. An overview on computational intelligence in control and decision making for complex systems will be given over the last four decades. Some real-world cases on power plant operation, information–driven safeguards, cost estimation under uncertainty for a large engineering project, and decision support for long-term options of energy policy will be illustrated for the potential use of computational intelligence related techniques in complex systems. Essential steps on implementing computational intelligence related techniques in industry will be presented via R&D, demonstration, and commercialization. Challenges and future research directions will be concluded in this talk.

As an adjunct professor of Shanghai University, Prof. Da Ruan gave us a lecture titled "Elementary guidelines for preparing an SCI journal article" on May 20, 2008. Many master students, Ph.D students and teachers were fascinated by his wonderful lecture. He answered many questions that were a concern for all researchers patiently and carefully, which obtained a warm applause. His lecture was very helpful for many researchers to write a high quality paper. He also gave us a lecture titled "Intelligent decision analysis support under various uncertain information in perception based complex systems" on July 16, 2010. In addition, he was also invited by Prof. Fei Minrui of Shanghai University to make some informal talks. He proposed many useful suggestions for our conference LSMS2010. He created many chances and brought some cooperation between Shanghai University, China and Hasselt University, Belgium. He did a lot of work and contribution for Shanghai University.

The abstract of the lecture "Elementary guidelines for preparing an SCI journal article" was as follows:

Advancement in science and engineering is frequently tied to publication of research in refereed journals (e.g., SCI indexed ones). Only with journal publication, your contribution to a field is no longer informal. The validity, the originality, and the fine-writing of your work are the key successful indicators for your future SCI journal publications. This lecture will guide you some elementary rules for preparing an SCI Journal article.

The abstract of the lecture "Intelligent decision analysis support under various uncertain information in perception based complex systems" is as follows:

Humans have a remarkable capability to perform a wide variety of complex decision tasks under various uncertain data and/or information based on perceptions. Decision support systems (DSS) with such perceptions often involve the use of traditional mathematical tools and modern artificial intelligent techniques. Due to the potential difficulties of dealing effectively with perception based risk assessment and management, information (or data) obtained by any means will be of very different nature. It may be heuristic or incomplete or data that is either of unknown origin or may be out of date or imprecise, or not fully reliable, or conflicting, and even overloaded. To allow an adequate interpretation of the information and to reach a conclusion by both traditional mathematical tools and modern artificial intelligent techniques, an integrated intelligent DSS that is able to deal with various uncertainties in real time is urgently needed. Hence, it is considered advantageous to have a sound and reliable mathematical framework available that provides a basis for synthesis across multidimensional information of varying quality, especially to deal with information that is not quantifiable due to its nature, and that is too complex and ill-defined, for which the traditional quantitative approach (e.g., the statistical approach) does not give an adequate answer.

I am doing post-doctoral research in Belgium. Prof. Da Ruan ever told me that we could go to Shanghai together to attend the ISKE2011 conference which was held on December 15-17, 2011. He was also invited by Prof. Fei Minrui to have a discussion when he was in Shanghai during the ISKE2011 conference. However,

Prof. Da Ruan suddenly passed away on July 31, 2011 from heart failure. His passing away made many schedules unfinished forever and left us too much regret. He made great contributions to the cooperation and communication between Belgium and China. We will remember Prof. Da Ruan as an outstanding scientist and an excellent Chinese.

3.16 Da Ruan in Fudan University: A Pioneer of Bridging the Academic Exchange between China and Belgium

Prof. Dr. Hanji Shang, Prof. Dr. Rongmin Li, and Prof. Dr. Yunmin Huang

Fudan University, Shanghai, China

In August, 2011, when we learned that Da Ruan passed away unexpectedly, we all felt tremendously shocked and deeply grieved. It means that an intimate friend, who had close cooperation with us for so many years, departed from us and Fudan University has lost an outstanding alumni.

Prof. Da Ruan entered into the department of mathematics, Fudan University for undergraduate study in 1979 and graduated in 1983. When he was a student here at Fudan, his excellent marks helped him earn appreciation from his teachers and his open and cheerful disposition won him trust from many friends. Prof. Da Ruan worked at Fudan as a teacher after graduation and then he was selected as a reserve force for an educational aid to Africa. Due to a change of plan, he went to Europe instead of Africa. Since 1987, Prof. Da Ruan took up advanced studies in two universities in Belgium, Ghent University and Leuven University respectively. After earning his PhD degree, Prof. Da Ruan remained in Belgium and gradually developed into an international influential Savant with great contribution in intelligence system and some other areas.

Although Prof. Da Ruan lived and worked in Europe, he always remembered his motherland, friends and Alma Mater. This was most evident in the last more than ten years, as he was devoted to promoting the academic exchange between China and Europe, especially Belgium. Prof. Da Ruan has visited Fudan University to give lectures for many times and he held various cooperative researches. With help of Prof. Da Ruan, teachers and graduate students of Fudan have the opportunity to establish good cooperative relationship with the European academic circles, especially that of Belgium. We can give many examples such as follows. In 2002, invited by Prof. Da Ruan, Hanji Shang, et al, attended the FLINS conference held at Ghent University and published their research papers. From 2005 to 2006, nominated by Prof. Da Ruan, Fudan University joined the China-Flemish Region cooperative project named "Data Mining and Information Processing: Methods and Applications", which is jointly initiated by Prof. Guoqing Chen (Tsinghua University, China), Prof. Geert Wets (Limburg University, Belgium) and Prof. Da Ruan. In 2007, a delegation from Hasselt University (whose predecessor is Limburg University), introduced by Prof. Da Ruan, visited Fudan University and signed a cooperation agreement between the School of Mathematical Sciences of Fudan University, China, and the Transportation Research Institute of Hasselt University, Belgium. In 2010, we had cooperation with Prof. Da Ruan and published our joint research journal paper.

During his visits in Fudan university, Prof. Da Ruan also established connection with the Suzhou Map2bit Software Technology Co., Ltd, which is a cooperation partner of Fudan University. Afterwards Prof. Da Ruan gave this company continuous instructions on its research work and their cooperative project was approved by the Suzhou government to be an international cooperative project of Suzhou, 2009.

During the 2010 Shanghai World Exposition, an event named “Belgium Day” was held in the Belgium exhibition hall, and several government officials of Belgium were in attendance. As one of the major activities of “Belgium Day”, Prof. Geert Wets from Hasselt University gave a report on the Belgian research results about transportation. As cooperative partners, professors and students from Fudan University attended this event. Together with us, Prof. Da Ruan made a plan for this event, which is probably one of the last academic activities he participated in China. After the Exposition, we continued cooperation with Prof. Da Ruan and made preparation for his next visit. Unfortunately, the sudden departure of him disrupted all these plans and left us with durable regret.

Despite that Prof. Da Ruan was departed from us, his kind voice and countenance as well as his important contribution to our Alma Mater will never fade away from our memory. May his soul rest peacefully in the Heaven! We also wish that the Sino—Belgium academic exchange, which was vigorously promoted by Prof. Da Ruan, could continue to exist and develop further, so that this bridging pioneer can be consoled and his unfulfilled wish can be realized.

Da Ruan and Hanji Shang in Fudan University, Shanghai, June 29, 2004.

Part IV

Thoughts by Da Ruan's Colleagues and Friends

4.1 Early Days in Ghent

Dr. Bart Cappelle[1] and Prof. Dr. Gert de Cooman[2]

[1] Royal Bank of Scotland
[2] Ghent University, Ghent, Belgium

Da arrived at Ghent University, as a PhD student on leave from Fudan University, in the autumn of 1987. He had been working mostly on numerical analysis at Fudan University, and his professors there had arranged for him to join our 'Seminar for Mathematical Analysis'. They may have been confused, as Da was often keen to mention, by this name for the department he was sent to, as they had told him he would be able to attend seminars on mathematical analysis there. Etienne Kerre, who was one of the people lecturing on that subject at the department, introduced Da to fuzzy set theory, which was then still a relatively young field, and also still relatively controversial. It was Etienne's idea to let Da investigate the properties of inference schemes, such as Modus Ponens and Modus Tollens, when generalized using Zadeh's approximate reasoning framework. Little or no advanced mathematical analysis was to come Da's way for at least a number of years.

Gert's Memories

I had also just joined the department, and was there to welcome Da on the morning he arrived, together with our head of department, Fernand Vanmassenhove, our teaching assistant, Jan Trangez, and the department secretary, Griet Pauwels. Da was the first Chinese person I'd ever spoken to, and I remember being surprised by his soft-spoken and self-effacing politeness.

We became good friends very quickly: it was very hard not to like Da. I tried to take care of him, who must have felt helpless and homesick, in the way of a young man of 22 who's never taken care of anybody before: clumsily and with the best of intentions, as best I could.

I remember teaching him the rudiments of Dutch, and getting basic Chinese in return; long walks in my home town; the mayonnaise I told him to add to the spaghetti bolognaise (a silly prank he always told people about but hopefully forgave me for); rambles at night through the beautiful streets of Ghent; long afternoons with Da and Bart, rich in coffee, discussion and banter in the Overpoort, our favorite student restaurant; his first Christmas eve with too much wine in Ghent at my grandfather's: alcohol usually made him very sick, and he got me really worried that night.

And then, during his PhD defense, there were the first signs, later to be so prominent, of a Da different from the shyly polite and usually defensive person we knew: we saw him full of confidence and enthusiasm, parrying silly criticism with a joke, turning his winning smile to his advantage.

After his defense, Da and Bart and I tried to keep in touch, but went our separate ways: life is what happens when you're not watching closely. And I wasn't watching closely enough. I saw Da on a few occasions, and then, the summer before last, I was privileged to celebrate his 50th birthday with him and his family (and with Etienne Kerre. Andrea De Kegel and Dirk Aeyels) in Shanghai. Da was the one taking care of us now, in an unfamiliar country, and he was so much better at it. That evening, we said goodbye on a busy Shanghai street corner. What makes us think we have all the time in the world?

Da was a good person as well as a good scientist. And I will always remember (and I am sure I am not the only one to have experienced this) how, every time he saw me again, his face and eyes began to radiate pleasure, and he gave me the feeling I was special to him. He was an exceptional man, and I consider it a privilege to have been his friend.

Gert and Da in the Great Wall in China 2010

Bart's Memories

Sometimes, people make decisions that change their lives years later, without realizing it the moment they take these decisions. Back in 1986, I decided to get introduced to the Theory of Fuzzy Sets. This was not a mandatory course but a choice for a field unknown to me. However, the modeling of uncertainty next to probability theory, triggered me somehow. Etienne Kerre introduced me in the matter in such an enthusiastic way that I really got excited to know more. Even better, I wanted to enter into research to find out things myself. Upon my graduation in 1988 and with Etienne's support, my little secret dream came true and I could enter the magical world of fuzzy sets.

That is how in September 1988, I joined the same Seminar, and met both Gert and Da. At that time, Da was working already for one year in the field of Fuzzy Logic and Gert entered the wonder world of fuzzy measures. Though I was focusing on Intelligent Data Reduction methods, linked to a project sponsored by a chemical plant in the Netherlands, Gert and Da introduced me to the fuzzy world they were working on. Da's research was extremely intriguing to me, and soon we ended up in long discussions in front of blackboards that changed into ample fields of formulas, assumptions, proofs and counterexamples. We became friends---that was inevitable, a deep and well-respected friendship was growing day by day. Our favorite waste of time ended up in what turned out to be Da's first scientific publication of a very long list. I did not realize at the time that I was taking part in a first major step in Da's career, nor fully understood what I was witnessing.

Our friendship was getting stronger. My passion for history and his eagerness to understand the small country he was living in, ended up in long discussions. Da was as well for me the first Chinese person I had ever met and China twenty-five years ago was quite a mystery. Hard to imagine a non-digital world without the Internet. He was teaching me about Chinese culture and history. The mysteries of China were little by little unveiled, while I was trying to explain Belgian history---if ever it would be possible for a foreigner to understand this small and complex country. Nevertheless, one day he wanted to see and know the region I grew up in. The country of my ancestors, destroyed and mutilated by the Great War, the land of John McCrae's poem, and the endless rows of crosses on the Commonwealth Cemeteries, the spaghetti in Ypres, where Da told me this time he would not have mayonnaise. The fields of destruction touched him deeply and we understood each other from that moment on, without talking to each other. It was the first of a number of visits, though not all in such dramatic scenery. I can only imagine what these visits meant to Da being thousands of kilometers away from his loved ones, staying with mine.

The cultural differences between us were enriching and sometimes led to very funny situations. Jokes in Chinese and Dutch were translated in such a way they were totally meaningless, but made us all laugh so hard that outsiders thought we had gone totally mad and urgently needed some days off.

As time went by, his PhD was emerging from numerous theorems, proofs and publications, substantiated on his desk by a pile of paper. On the one hand, I was happy to see his PhD grow, on the other, the hour our paths would separate came closer. As with anything he did, his PhD was getting so close to perfectionism and he wanted to be fully prepared for the ultimate defense. Gert and I trained him, where I was obviously the nasty guy and Gert challenged his scientific work. Just like Gert, I saw Da grow in his role, day by day during the weeks of preparation. It must have marked a milestone in his life---I realized later on. His exposé I still remember as if it were yesterday. Assertive and firm in what he was explaining, brief and to the point, he answered any question he got. No one would dare to challenge his dissertation. Gert and I were both amazed and proud.

Da left Ghent for the SCK. We stayed in touch for a long time, though the last couple of years occasionally an exchange of e-mails how busy we both were. Time flew, and in the beginning of last July I had an idea that clearly needed further development. I gathered a lot of information and was convinced that Da's work could serve as a basis for an application outside the nuclear world. During my holidays, later that month, I started to prepare for the old eighties gang to reunite. I compiled a massive number of wild ideas, an excellent excuse to meet and build again together. One e-mail broke me a couple of days later. We did reconvene ... and we did write a paper, not with you but about you, Da.

Da and Gert have taught me more about true friendship than anyone else has. Indeed, my dear friend, it was a privilege to have known you.

4.2 Da, a Good Supervisor and a Nice Friend

Dr. Yun Shi

Atlas Copco Airpower, Antwerp-Wilrijk , Belgium

The first time when I met him, I could not believe that he was Da Ruan who had already been a leading scientist in the area of fuzzy mathematics and fuzzy control in nuclear research, because he looked so young. Before I came to Belgium to start my PhD study in Ghent University, Da, one of my future supervisors had sent me some books he had written together with Prof. Dr. Etienne Kerre, the other supervisor of me. In one of the books I have read about a short profile of Da and Etienne. According to their experiences and achievements in the research world, they were both men having quite ages with white hair and long beard in my imagination. That is why it was a total surprise when I found out that Da was still quite young: he was forty-four years old when he became my supervisor six years ago.

Both Da and Etienne are very good supervisors. My main working place was in Ghent while Da worked in the Belgium Nuclear Research Center in Mol which is about 100 kilometers far away from Ghent. From time to time Da came to Ghent to sit together with Etienne and me, talking about the progress of our research work. Moreover, he always sent me the most pioneering research works about fuzzy control and fuzzy decision making in the literature to help me gain a meaningful study direction. Sometimes I went to Mol to visit him. There he had not only explained me how did fuzzy control work in nuclear research but also told me how had he grown up in the research world. He said that as a foreigner one had to pay double or even three times effort to achieve the same results as a local person. He encouraged me to do my best to face and overcome all the difficulties.

In 2007 when I participated EUSFLAT international conference in Ostrava I was so nervous before my speech when I noticed that the audience was full of top experts in the area of fuzzy mathematics. Da asked me to be calm. He told me to give the speech clearly and that the experts would give me nice suggestions afterwards. It was just like what he had said. The remarks and questions I received after my speech gave me a lot of ideas of further study. Moreover, the nice impression of the experience of this speech which was one of my very first speeches stays a very good part of memory of me, with the important encouragement of Da and Etienne.

Da had students all over the world. From time to time many master or PhD students from Belgium, China, Italy, India and Turkey came to Mol for an internship. He always introduced us to each other so that we could have possible cooperations in the future. Our research group in Ghent University had from time to time visitors from another university or another country. Da always asked them also to pay a visit to Mol. For those who for the first time came to Belgium Da did not hesitate to help them, to find them an accommodation place, to show them around the cities (Ghent, Mol, and so on).

Da loved his wife and sons very much. I can always remember how proud he was when he mentioned his wife Greet and his two sons Alexander and Nicolas. He said that Greet sang beautiful songs, he said that Alexander and Nicolas were always so active that they did not feel tired even if they ran dozens times around a playground, he said that they spent nice time when they travelled together to China…

I think that Da turned to a star in the universe who is still looking at his family, his friends and his students. When we look at the sky in the night we will always think of him.

4.3 Untold Thoughts

Dr. Lusine Mkrtchyan

Belgian Nuclear Research Center SCK·CEN, Mol, Belgium

I got to know Da in summer 2009 when he was one of the invited speakers in summer school of Advanced Artificial Intelligence in Northern Ireland. After the summer school I asked him if I could join his research group in Belgian Nuclear Center SCK · CEN. I spent almost a half year working with him on developing a new type of fuzzy cognitive maps in Mol. At the end of my stay he suggested to write a proposal for post doc position in SCK for a project of Safety Culture Assessment.

I remember the day when he called me to tell that our proposal succeeded and we will start to work from September 2011. We even were supposed to share the same office. Unfortunately, I started my work without him, and when I entered his office and saw his notes about meetings, I had a feeling that he will just return now and continue to work…

One of my most touching memories of Da is his gift of a Chinese Chess Set. I told him that I am learning Chinese Chess and that I started to like it even more than European Chess, and in our next meeting he gave me his own chess set telling me that he will buy for him another one in China. I will treasure this small chess set forever…

Another nice memory of him is his invitation to Chinese New Year. I was one of the few non-Chinese participants of the evening, and I enjoyed the evening very much. There was also his wife and his two sons. He cooked a very delicious dish for this event. Then I followed his game of mahjong with other Chinese people, and I promised him to learn it until the next meeting of SCK Chinese community that was supposed to be this October for Chinese Moon Festival.

And I shall never forget our last meeting in Hasselt in June 20, when he just came back from Shanghai. We discussed summer holidays, our work stages for the autumn and I could never imagine that it was the last time that I could see him.

He was able to give so much positive energy, so much optimism, so that after each of our discussion about our work I was motivated endlessly. With his way of treating his job, with his unlimited devotion, with sincere enthusiasm he was and will be always the example of a great scientist for me.

I will miss you always. I met you in the beginning of my career, and during all my researcher activities I will always be thankful to you for your big contribution to my professional development…

Thank you my dear Professor Da Ruan.

4.4 Da's Last Initiative: Launching the SSRAOC Workshop

Ir. Benny Carlé

SCK•CEN, The Belgian Nuclear Research Institute, Mol, Belgium

Missing Da, and still somewhat in shock realizing he will never take us in tow again for adventures like this - starting up a new research community by organizing a workshop, setting up collaborations and organizing scientific publications in special issues and books - we had the SSRAOC workshop as planned, in Antwerp on January 29th-$31^{st,}$, 2012.

The workshop started with an honorary session, in which we shared memories of collaborations and friendship. Ashok Deshpande performed a classical Indian song "*Seva Dharma utam [service to all is the religion,,]*" and Greet Van Tichelen, Da's wife, sang the beautiful songs "*Ich liebe dich*", "*You are the new day*" and "*In this heart*", together with two of her friends. Jie Lu gave a presentation "Da Ruan: friendship and leadership". Eric van Walle talked his experience to work with Da and particularly their conversation just before Da passing away. These were unforgettable moments. During the preparation of the workshop, I met Greet. This made me realize that Da was not only an extraordinary professional, but that he also had a wonderful wife, and a magnificent family.

The workshop, on Safety and Security Risk Assessment and Organizational Cultures, convened 45 researchers and experts. Five keynote speakers presented their views on the emerging field. And we had 20 state-of-the art presentations on current practices and new research. Moreover, round table discussions, and many intensive discussions during coffee breaks, lunch, dinner and late night drinks in the hotel bar made the conference into an unforgettable experience. The workshop was full of energy, the quality of the contributions was high, and the discussions were lively and constructive. We all left with a feeling "...let's do this again".

Da was a visionary, and this workshop was a perfect illustration of this. The focus of the workshop was on how to assess safety culture and on how to deal with uncertainty in the safety risk assessment. Da was imaginative enough to see that new developments in mathematics would help us tackle challenges with safety in high-reliability organizations that work with complex technological installations.

Nuclear organizations have had a long tradition on emphasizing safety. In the last decade we have become more aware of the importance of safeguards and security. This has led to the implementation of similar policies to security as to safety, although approaches specific to security remain in place as well. It was Da's enthusiasm bringing together a team in SCK•CEN with expertise in safety and broader domains of security and safeguards: Frank Hardeman, Fernand Vermeersch, Catrinel Turcanu, Klaas Van der Meer, Sabine Janssens and myself.

And Lusine Mkrtchyan's project on safety culture assessment, starting in fall 2011, fitted in perfectly in the scope. When Da came with the idea, late 2010, we all wanted to contribute to this workshop. Most of us had done projects where Da was our mathematics partner: projects on energy policy, waste disposal policies, on uncertainty in design, on open source information for safeguards, etc. The workshop was in line with the work we had been doing with Da for nearly 20 years. We saw the SSRAOC workshop as an opportunity to better understand, assess, model the influence of human factors and organizational, cultural elements in our safety and security studies. We discussed safety culture, security culture, safeguards culture and how they blend together in an organizational culture; how to approach security and safeguards by design.

When Da passed away so suddenly last July, most of the organizational work for the workshop had been completed. Da had taken the lion share of it. We had invited speakers, briefed journal editors, and published a call for papers out. All this was facilitated by Da's legendary network, that he had built over the years, and that was based on outstanding scientific competence, and his warm personality. So many people from all over the world were happy to work with him. Da always generously shared his network with us, his colleagues. Now that Da has passed away, we sincerely hope that the young scientists from his networks will continue to come and work on SCK•CEN projects.

So the workshop preparation was typically Da, hardworking and always going for the best possible result. Never satisfied with mediocre quality, always going for top-notch. When discussing the best place for this workshop about a year ago, we chose for Antwerp, the hotel near the railway station, (Astrid – Radisson Blue) and with the official workshop dinner in the nearby Antwerp Zoo. Some of us remembered the 3rd FLINS conference in 1998 in the same setting, and we had good recollections of that event. The logistics for the workshop were taken care of by Els Van Musscher, our administrative jack-of-all-trades. During our workshop, it felt good we could have the honorary session in a place linked to Da. And Greet told us Da's Chinese food shop was just around the corner from the hotel....

The workshop brought many friends from Da back to Antwerp: Etienne Kerre, Jie Lu, Ashok Deshpande, Georges Van Goethem, Enrico Zio, Ahmet Can Kutlu and so many from our research center SCK•CEN. We talked about Da, about our personal relations with Da, about Da as a family man, as a friend, as an inspiring person in all our lives. And we discussed science, our research, our work, our passion. We withal felt the urge to preserve this community. I hope that is a tribute to Da, a contribution to his heritage.

I miss Da. I arrived in 1994 in SCK•CEN in Mol as IT-manager, and Da was already an outstanding researcher, doing advanced projects, always delivering quality that was internationally recognized. For 17 years, Da has been a role model for me in SCK•CEN. Our first collaboration was on building a demo system on the fuzzy controller for the BR1 reactor. The years following, we worked on small projects together; often fuzzy set extensions on multi criteria decision support, which we applied for so many applications in the DSR (Decision & Strategy research unit) and SPS (Society and Policy Support) expertise groups in SCK•CEN. We made many plans to start a bigger project together, but they remained vague. The moment was never completely right, the funding missing and it always seemed like there would be a lot of time left to do it. I missed out on the opportunity to work on a bigger project with Da, but I hope my contribution to the SSRAOC workshop will compensate it.

4.5 A Personal Tribute to Our Dedicated Friend Da Ruan

Prof. Dr. Madan M. Gupta

Intelligent Systems Research Laboratory, University of Saskatchewan,
Saskatoon, Canada

Dear Da:

It was on Monday, August 1st, 2011, that I received from our friend Prof. Tianrui Li the following heart-breaking E-mail message:

'Our best friend Prof. Da Ruan (Chair of FLINS and ISKE) suddenly passed away yesterday (Sunday, July, 31st, 2011)'

My young charming friend Da, first I did not believe it - how it can be possible with you because only a few days ago I received an E-mail from you, but subsequent phone calls and E-mail messages confirmed that this 'impossibility' is true!! Sadly, yes, our friend Da you have left this earthly home for your heavenly abode and we do not have any choice but accept this reality.

Da, perhaps you will recall our *first meeting* in July 1996 in San Francisco at the 13th IFAC World Congress (June 30 –July 5, 1996) and the *last one* this year again in San Francisco during the World Conference on Soft-Computing (May 23-26, 2011). During our first meeting you introduced to me your newly founded international research forum on Fuzzy Logic and Intelligent Technologies in Nuclear Science (FLINS) and the conferences under the umbrella of FLINS. Since then I have had the opportunities to actively participate in six FLINS (*1998 Antwerp, 2000-Bruges, 2002-Gent, 2004-Blankenberge, 2006-Genova and 2010 - Chengdu*) for presenting my research work and tutorials in the field of fuzzy logic and neural networks. These FLINS conferences gave me an opportunity to renew friendship with my many old friends, such as Prof. Dr. Etienne E. Kerre, Gent, Belgium, and make many new ones. During these visits to Belgium, I have had the very delightful opportunity of meet with your charming wife Greet, and two talented young sons Alexander and Nicolas as well.

Dear Da:
In your smile,
we saw a glimpse of Applied Computational Intelligence!
And your laugher
used to warm our hearts with fuzzy joy!
Still you are in our hearts and
will remain so for ever.

You were the light of Greet's life, your charming wife,
and two bright sons Alexander and Nicolas;
I am sure they will always keep their path shining,
the path of intelligence that you have laid down
for them over the years.

You became the life guide for nuclear reactors
through the mathematics of 'intelligent control' and
the language of 'fuzzy logic!'

Da, you were the driving force behind the founding of
'The International Journal of Computational Intelligence Systems (IJCIS)', and
Fuzzy Logic and Intelligent Technologies
in Nuclear Science (FLINS) and,
for your innovations in these fields,
the international research community will always remember you.

Da, having a, friend like you
always brought a special joyous feeling in our breath!
May the pleasure and joy of 'computational intelligence'
that you have left with us
may continue to be a driving force for the generations to come!

Da, now you are in your heavenly abode,
but the seeds of 'intelligence in nuclear industry'
that you have planted, will always give us some nourishing fruits.
Da, your smiles & teachings will always remain
with us in our hearts and neurons!
Dear Da, your fuzzy work will always give us some strength
for the betterment of this technological world.
In conclusion, Da, goodbye to your physical existence but
not to your everlasting soul, smiles & teachings!!

Your friend & admirer:

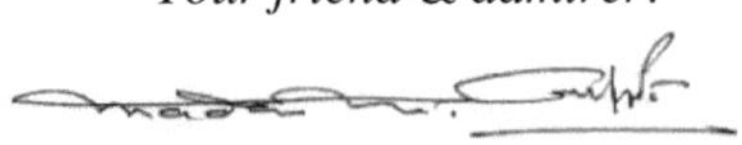

4.6 Da as a Flawless Organizer

Prof. Dr. Witold Pedrycz

Dept. of Electrical and Computer Engineering, University of Alberta, Edmonton, Canada

For the first time, I met Da more than a decade ago at some conferences on fuzzy sets and their applications. Since then we had a lot of interactions, mainly through emails when discussing some concepts, exchanging ideas, and talking about some research projects. I greatly appreciated his tremendous energy, dedication, research vision, and extraordinary abilities to connect with others.

I can vividly recall FLINS conferences – truly outstanding gatherings with well – defined, timely, and important themes. These conferences formed a vibrant environment and fostered truly collegial and thought-provoking venues for innovative research. This was Da's very idea: to create a unique form of conferences that bridge theory and remarkable down-to the earth and challenging applications such as those encountered in nuclear industry. I enjoyed the venues of these conferences and had an opportunity to appreciate their flawless organization and a great deal of opportunities offered to young researchers. Da was behind these unquestionable successes the conferences enjoyed. Fuzzy risk analysis was one of his leading research areas: he set up the long-term agenda and persistently realized it followed by others in the community.

We interacted on several publishing projects. Da served as a Guest Editor of a special issue published in *Information Sciences* (Elsevier). It was a genuine pleasure working with him. At that time, I learned about his other qualities that made this special issue an undisputable success: prudent planning and organization, careful arrangements of reviews and a constructive, timely feedback provided to authors. The flawless organization of the special issue has made it a remarkable success.

Definitely, Da will remain in our memories as a truly outstanding colleague, creative and innovative researcher, leader, and truly exceptional human being.

Witold received a FLINS award from Da in FLINS2008

4.7 Da Ruan, Our School Classmate and Best Friend

Qingyu Zhu and Xiaoling Xu

Shanghai, China

On July 31, 2011, Da Ruan, our best classmate, suddenly left us and went to Heaven. So far, we still could not accept the fact. Da, have you really left us? Are you happy in Heaven? We miss you!

We still remember you, Da, in our high-school-age in Changzhi High School in Shanghai: in the family, you were an obedient son, and a so kindly elder brother and a respectful friend to your brothers and sisters; in the school, you were so excellent in every aspect, our classmates modelled on you and our mentors felt proud of you.

We still remember you, Da, when you settled down abroad: you were the beloved husband to your wife and the kindly father of your loved sons; and you had a flourishing career with your hard work.

We still remember: you always gave your time to your classmates for sharing your working experience and life thoughts once you have any chance to visit your motherland no matter how busy you were.

You were really a good teacher and a helpful friend of us.

Although you have gone to Heaven, you left use your voice and face forever!

Da, we will miss you forever! May you rest in peace! (Translated by Jun Ma)

悼念好友——阮达

朱庆玉，徐晓玲

2011年7月31日我们的好朋友，好同学阮达，突然离开了我们，去了天国。至今为止，我们还是难以接受。阮达你真的去了吗？你在天国过的好吗？

我们非常非常想念你。

记得在中学时代，你是一个孝顺的儿子，同时在妹妹弟弟眼里，你又是一个多么慈祥的兄长和朋友。在学校里你又是一个德智体全面发展的优秀生。是我们同学中的楷模，老师的骄傲。

后来，你定居在国外，成了妻子眼中的好丈夫，孩子心中的好父亲。同时你的事业也随着你的成长不断的蒸蒸日上。更可贵的是，你每次回国演讲，虽然时间很短，可你总是不忘在百忙中抽出一点时间和我们同学聚会，关心我们的工作和生活，是我们的良师和益友。

虽然，今天你离开了我们，但你的应容相貌将永远留在我们的心中。

阮达——我们永远怀念你

祝你在天国快乐！

4.8 The Memories That Long Last

Prof. Dr. Guoqing Chen

School of Economics and Management, Tsinghua University, Beijing , China

My last meeting with Da Ruan was in Spring 2011 at Tsinghua University, where we were talking about his plan for the next visit to us in Fall 2011 for a bit longer stay on campus to have more chats and thoughts on some future research ideas. As a matter of fact, this last meeting had been one of many gatherings with him during 23 years since we first met in 1988 when I visited Professor Etienne Kerre in Gent right prior to my doctoral studies at Catholic University of Leuven (KUL).

The heartfelt shock of Da's passing away made me to recall many memories...... his passion, hardworking and heartedness, as well as the achievements in the fields. It is hard to express our sorrow feelings in words, and what I could write at this juncture is a collection of a few pieces of memories that will long last.

Da and Guoqing is SCK in 2004

The professional life of Da could be seen by many of us as to be of a passion nature. Over years his strong spirit and devotion in pursuit of his career and desire impressed me greatly, from our early discussions at his dorm in Home-Astrid of RUGhent for his FIO (fuzzy implication operators) exploration and its future prospects in his doctoral research, to his presentations facing a group of Chinese scholars in Belgium for launching the FLINS symposium series with high excitement, to his enthusiasm in applying fuzzy logic to control systems at SCK-CEN, and to his frequent contacts and visits with numerous Chinese scholars in China and overseas for various activities and events. More noteworthy was his continuous effort and sense to proceed ahead for novel ideas and initiatives.

Moreover, hardworking is a characteristic of Da, which I believe many would agree with. For nearly two decades, Da had been engaged in many proceedings, special issues and edited volumes, where he was fully committed in all aspects of the publication process, not only in content but also in format. I remembered for several times he mentioned to me that he had worked overnight in Gent, Mol, China, etc. at home, office and/or hotels for editorial handling and even typesetting tasks. We often chatted, sometimes with Etienne, for how and when we could slow down the pace of working, which we had found to be an extremely difficult goal to reach, if not impossible.

Furthermore, Da had a heartedness personality. He had been very sincere in helping people. Since mid 1990's, he had started to host young scholars such as post-docs and visitors at SCK-CEN, of which many were Chinese researchers. In recent years, he visited a number of Chinese universities and provided a great amount of guidance and advice to the students and faculty in support of their research and disciplinary development. In my previous international cooperative projects (namely the China-Flanders Cooperative Program) with Ghent University and Hasselt University, where Professors Kerre and Wets were respective principal investigators on the Belgian side, Da was involved in hosting and guiding, in Gent or Mol, some of the Chinese scholars who paid their academic visits under the projects. In addition, his heartedness was also evidenced, as an example, in the FLINS conference series where he gathered a large group of international scholars, many of which are world-renowned ones, in a way that well combined high-quality research with close relationship.

Importantly, having the passion, hardworking and heartedness, coupled with other talents, his academic achievements were outstanding. In addition to the contributions to fuzzy reasoning and applications, his research had been extended to a broader realm, such as intelligent techniques, computational intelligence, risk management, traffic decision making, etc., resulting in a significant amount of impactful outcomes in many ways that range from publications, to invited speeches, to roles on journals' editorial boards, and to chair-ship in international conventions. It is worth mentioning that the FLINS conferences were well received by the fuzzy logic and related communities, in which he was the leading figure, and that the ISKE (International Symposium on Knowledge Engineering) conferences, originated from the China-Flanders Cooperative Program, had been successful in previous years as a series, to which Da contributed greatly in the course of its development. In particular, Da devoted considerably his energy to the new journal IJCIS (International Journal of Computational Intelligent Systems) as the Editor in Chief during the past years, with remarkable progress made in its well-being.

Hereby in this memorial episode, what I could conclude now briefly is that we have a reflection on how a scholar like Da strives to achieve his goals in the whole career, which others respect.

4.9 Eleven Meetings with Prof. Da Ruan

Dr. Xiaohong Liu

Management College, Southwest University for Nationalities, Chengdu, China

From 2002 to 2010, I have met respected Prof. Da Ruan for 11 times in China, Belgium and France. Until now, the 11 times experiences of meeting Prof. Da Ruan are still vivid in my mind.

In the summer of 2002, I made acquaintance of Prof. Da Ruan in Southwest Jiaotong University, Chengdu, China. At that time, I was doing PhD research at Intelligent Control Development Centre, supervised by Prof. Yang Xu, just facing the challenges of writing and publishing academic papers. Prof. Da Ruan gave us lectures on how to write academic papers to help us get into state quickly. The second time I met Prof. Da Ruan during the spring of 2003 in Beijing, at an international conference in "Fuzzy Information Processing Theory and Applications". Prof. Da Ruan not only gave me a very suggestive guidance, but also guided the other young people attending the conference carefully. In the winter of 2003, I met Prof. Da Ruan again when I was attending the international conference "2003 China-France Conference on Sensory Evaluation with Smart Technology'' in Shanghai, China, along with my PhD supervisor Prof. Yang Xu and my co-mentor Prof. Xianyi Zeng in France during my postdoctoral period. At that conference, I learned about the unique concept of intelligent information processing and Prof. Da Ruan's great consideration of young students. It was the third time when I met Prof. Da Ruan. When I was in my postdoctoral period, Prof. Xianyi Zeng, Prof. Da Ruan and I made a further discussion on a conceptual model of decision-making at Belgian Nuclear Research Center (SCK). In 2005, the international conference IMACS2005 was held in Paris, France. Because there were many Chinese scholars participating in the conference, Prof. Da Ruan has taken the role of translator and helped them quite a lot in communication with others. It was the fifth chance for me to meet up with Prof. Da Ruan. The sixth time I saw Prof. Da Ruan was when I visited his house in Mol, Belgium in 2005 and had a taste of delicious food he cooked to which I was very impressed by his cooking skills and his kindness. The next year, Prof. Da Ruan accepted my invitation to be a visiting professor and made a wonderful guest lecture in Southwest University for Nationalities where I was working. In 2007, I met Prof. Da Ruan again for the eighth time. At that time, he was the general chair for the international conference on "Smart Systems and Knowledge Engineering'' held in Chengdu, China. His great efforts have made that conference very successful.

After the Wenchuan earthquake, it's the ninth time I met him in May 2008. The world is shocked by the huge casualties caused by the earthquake. According to the official announcement, the number of the deaths and disappearances was more than 80000. Now I still remembered that Prof. Da Ruan visited Chengdu according to the work plan, even though the earthquake just happened a few days before. Influenced by the air traffic control and earthquake relief work, he could not do the teaching and research in the Southwest Jiaotong University and other places as usual. Then Prof. Yang Xu (being the Deputy Director of Sichuan Provincial Information Industry Department and being responsible for Supporting the Shuangliu International Airport Air Traffic) together with his students welcomed Prof. Da Ruan in a farm-house around the Shuangliu International Airport, and both of them had a deep discussion about earthquake relief and post-earthquake reconstruction. Even though the aftershocks and the situation were still unsafe, he insisted in working regardless of personal safety and did his best to support China's Sichuan earthquake relief. The next time we met each other was in Chengdu again in 2009, when he made an effort to help the cooperation between Southwest Jiaotong University, Southwest University for Nationalities and the universities in Belgium. At the same time, he proposed a professional report for the MBA students in Southwest University for Nationalities. The last time I met him is on the 9th International FLINS Conference on Foundations and Applications of Computational Intelligence (FLINS2010) held in Sichuan in 2010. The academic conference was created by him and had great international influence. As the general chair, he held a high-level academic conference successfully.

Memorizing our beloved Prof. Da Ruan, we can't believe his sleeping forever. At the same time, his pursuit of truth, serious attitude toward work, emotional ties to the spirit of our compatriots and overseas Chinese scientists made a good example for us. He will be missed and memories about him cherished forever!

Da lecturing in Southwest University for Nationalities in Chengdu in 2006

4.10 To Da Ruan, Our Friend Forever

Prof. Dr. Liangzhong Yi and Prof. Dr. Zheng Pei

School of Mathematics and Computer Science, Xihua University, Chengdu, China

Our academic cooperation with Professor Da Ruan began in 2004, when our mentor Prof. Yang Xu from Southwest Jiaotong University asked Professor Da Ruan to help us modify our research paper "Extracting a complex linguistic data summary from a personnel database via a linguistic aggregation of simple ones" which was submitted to Information Sciences. Professor Da Ruan accepted the invitation and helped to modify the paper word by word, meanwhile he also provided a lot of valuable advices. With his help and advices, the paper was quickly accepted by Information Sciences in a Special Issue. This is our first English scientific journal paper. Professor Da Ruan's knowledge and enthusiasm deeply impressed us. In 2007, we co-authored "Handling linguistic information over the internet based on a multi-agent system" in the International Journal of Intelligent Systems, where he put much effort to improve the quality of the paper in terms of writing and technical aspects and the paper was also published soon after. He has helped us to improve our English writing skills for scientific papers many times.

In 2006, Professor Da Ruan was appointed as Guest Professor at Xihua University when he came to our university for guest lectures. He deliberately left out the time to discuss with core researchers of our School of Mathematics and Computer Science to share the experience and skills in academic papers writing, research faculty-building and other issues. At the same time, he exchanged experience with the editor of the Journal of Xihua University on how to enhance the academic level of the journal. At this point, we built up strong cooperation relationship with Professor Da Ruan. On May 12, 2008, Sichuan Wenchuan earthquake happened, the first e-mail we received was sent by Professor Da Ruan. Just after the earthquake, Professor Da Ruan went to Chengdu to visit us and brought us great care and his support from different aspects which indeed made us feel much warm and grateful. In addition, every time of his visit, he brought us Belgian chocolates, we can feel his care and he is such a nice person with great kindness!

Mediated by Professor Xu Yang, in 2009 we published a monograph with Professor Da Ruan. For Xihua University, this is the first monograph in English. For this monograph, Professor Da Ruan invited me (Zheng Pei) to visit SCK in Belgium. During this one month visit, we discussed the manuscript's content and determined the framework of the manuscript. Professor Da Ruan polished the manuscript further in great details. Our monograph "Linguistic Values based Intelligent Information Processing: Theory, Methods, and Application" was published in December 2009 by Atlantis Press & World Scientific. In May 2009,

Belgium was slightly cool, but my heart was warm. Professor Da Ruan arranged my accommodation and welcomed me by nice Chinese food at home at the first day. Conventionally (all Chinese friends who came to Belgium, Professor Da Ruan have always invited them to his house for a meal). I went to Professor Da Ruan's house, saw the vegetables, fruit trees and the caged chickens in the yard. Indeed as Prof. Jie Lu from the University of Technology Sydney, Australia said, Professor Da Ruan's home is simply a farm yard, he was such a person who was both a knowledgeable scholar and who knew how to enjoy the beauty of life.

When we were looking forward to further collaboration with Professor Da Ruan, he suddenly passed away and left us, like a bright meteor in the sky to illuminate our hearts at the same time left us endless regret! When looked up at the sky, we always think which star will be Professor Da Ruan?

4.11 To Deeply Cherish the Memory of Da Ruan, My Friend since 1988

Prof. Dr. Xuzhu Wang

Department of Mathematics, Taiyuan University of Technology, Taiyuan, Shanxi, China

Da Ruan passed away suddenly, left us not only with the enormous sadness and shock but also the endless recollections of the past.

I was sent to Belgium by Shanxi provincial government as a visiting scholar in 1988. Considering my research area is related to mathematics of fuzziness, in 1991 I contacted Professor Etienne Kerre from Ghent University for cooperation and his guidance and was finally accepted by him. In the follow-up communications with Etienne, I heard of Dr. Da Ruan for the first time (he has obtained the PhD degree from Ghent University). I got in touch with Da soon who told me much information about Belgium and the life abroad. From Da, I also obtained some help and encouragement which have been much important for my starting of research and life in Belgium. I arrived at Ghent University on Nov. 17, 1993 and started my study and research life as well as the face to face contact with Da since then.

Soon after I was in Ghent, Da invited me to Mol for a short stay (about one week), while we discussed AHP-related issues and revised a paper entitled "Fuzzy Analytic Hierarchy Process" which was published later on in the International Journal of Uncertainty, Fuzziness and Knowledge-Based Systems. During the stay, he expressed the wish that we could carry out a further cooperation in Mol if possible. In the autumn of 1994, my financial support was approaching the end. Approved by Etienne, Da offered me a position as a trainee researcher in Belgian Nuclear Research Centre in Mol (SCK-CEN). From then on, I cooperated with Da and Bartel *et al.* in Mol for 15 months, while we spent much time together and have been getting along well with each other. As a result I knew Da better than ever and the friendship between us 1996 began to establish and grew with time. During this period, I participated in his research project "Fuzzy Logic and Intelligent Techniques in Nuclear Science" and published two joint papers with Da. Thanks to Da's encouragement I took part in the first FLINS conference organized by Da. It is my first time in my life to present my research results in English in an international conference. In the conference, I also met some famous scientists in the research community whose names I only saw from the books or journals. It was one of the most unforgettable experiences in my life.

In addition to research connections, Da was also concerned and connected with my life. We often had dinners together, and played cards or table-tennis during weekends. Da was famous among his friends for his cooking skills and cards-playing techniques. My stay in Mol was pleasant partly owing to Da's help.

In the beginning of 1996, I returned to Ghent University, and continued my PhD research work. Although the contact with Da became relatively less, we met and had a chat every time he came to Ghent. In March, 1997, I defended my PhD thesis successfully and Da was one of committee members for my defense. It is no exaggeration to say that Da's help had been accompanying me for every stage from my preparation for going to Belgium to the end of my study and research in Belgium.

After I left Belgium on June 30, 1997, we had e-mail contacts from time to time. However the cooperation has not ended completely. In 2009, Da, Etienne and I jointly completed a book entitled "Mathematics of Fuzziness – Basic Issues" which has been published by Springer. Da made a lot of efforts in order to get it published in good quality. In the same year, Da came to Taiyuan in China for a potential cooperation with Taiyuan Radiation Protection Institute and we had a short meeting. Then in 2010, Da attended an international conference in Beijing, and I went to Beijing to meet him. I cannot believe that that meeting has been a goodbye to Da forever.

Finally, I would like to speak in Chinese: 安息吧, 达!

4.12 Recalling Professor Da Ruan

Dr. Li Zou

School of Computer and Information Technology, Liaoning Normal University, Dalian, China

Professor Da Ruan passed away unexpectedly. I received the news from an email when I was about to send a manuscript to him. I could not believe my eyes and tears streamed down my face. How desperately I hoped that it was not true!

Prof. Da Ruan and I had not met each other until the ISKE hosted by Southwest Jiaotong University in 2007. We had been communicating by email before our first meeting there and he had always been patient and serious about consulting and seeking help of a doctoral student like me. We met face to face when he came to register while I was a volunteer for the ISKE. He was so energetic and with such a good memory that he mentioned to me that I was one of the doctoral students of Southwest Jiaotong University and that he remembered what we had discussed in our emails.

Professor Da Ruan came to lecture in Chengdu many times and each time he explained the profound in simple terms and guided a practical and definite way for out scientific research. Professor Da Ruan was a devoted and serious scholar. He often made international long-distance call from Belgium to discuss some academic issues with me whenever I consulted him so as that some of my research was inspired by him. He revised every of my papers carefully and discussed them with me and he often emailed some references and academic information to me. He always encouraged me with such remarks as "You're intelligent!" or "Your English is quite good!" and his encouragement indeed made me more confident and work harder. He always conveyed tirelessly his ideas and experience to others with all his heart. I learned so much from him that I was not only broadened in vision but also influenced by a noble soul!

Professor Da Ruan was approachable and humorous and all of us felt relaxed and delighted every time we were chatting or having dinner together. Professor Da Ruan once encouraged me to host an international meeting and gave me many advices about it while he, Professor Tianrui Li and Professsor Guangquan Zhang and I were drinking tea together after the FLINS2010 last August. And we discussed it in detail for a long time pleasantly. I knew from him that we do our research work only for scientific interesting. It was such a pity that it was my last meeting with him!

I admired his great academic achievement, noble personality and great character. I will bear him in mind forever. Professor Da Ruan has gone but everything about him has engraved on my mind. He was my good teacher, my old brother and my good friend. He influenced and benefited all my life. I am deeply moved by and deeply grateful for all this! All that I could do is expressing my best wish for him.

Prof. Da Ruan and Li in Chengdu, China after his tutorial in 2008

4.13 Prof. Da Ruan in My Memory

Dr. Yan Yang

School of Information Science and Technology, Chengdu, China

Suddenly heard the passing of well-known scientist Prof. Da Ruan, I was sadden deeply! His voice and expression is still fresh in our memory. I will miss him forever. In my memory, Prof. Da Ruan is a kind and friendly person. He is open-minded and sincere with many friends all over the world. He is respected and loved by friends. He worked hard and obtained fruitful academic achievement, however, he was always approachable, affable, and indefatigable.

In the end of 2008, it is my first time to meet Prof. Da Ruan. He was invited by Prof. Yang Xu and Prof. Tianrui Li to visit School of Information Science and Technology, Southwest Jiaotong University, China. He gave a report entitled "Computational Intelligence in Complex Decision Systems" to teachers and students.

Prof. Da Ruan reported at School of Information Science and Technology, Southwest Jiaotong University, Chengdu, China, in 2008

His report was wonderful. His vivid language led us to explore the huge space of computational intelligence and complex decision making systems. I was given a deep impression that his passionate and humorous speech made the lecture hall, accommodating 200 persons at the same time, packed with people.

After that, I met him many times when he came to our university for academic exchanges or I attended International conferences he chaired, including ISKE2009 (Hasselt, Belgium), MDMKD2009 (Chengdu, China), FLINS2010 (Emei, China), ISKE2010 (Hangzhou, China). The last time I met him was in the end of last year. I remembered that there are only two days left for the New Year. He abandoned his holidays and was away from his family to visit our university to guide PhD students, attend a conference in Chengdu and report his latest research work in our lab.

Therefore, we kept close in touch by emails and then I had a chance to cooperate with him. My cooperation with him was a research paper on consensus clustering in intelligent decision support systems submitted to International Journal of Knowledge-Based Systems. Prof. Da Ruan gave me many valuable suggestions and revised this paper carefully more than ten times not only on the formulae and expression of the paper but also the punctuation marks. Each time I received the revised version of the paper from him, I was impressed by his rigorousness.

Goodbye, Prof. Da Ruan. You will remain in my memory forever.

4.14 Scientific Heritage of Prof. Da Ruan in the University of Ostrava – Life Lessons in Our Academic Career

Prof. Dr. Irina Perfilieva and Prof. Dr. Vilém Novák

Centre of Excellence IT4Innovations, division of the University of Ostrava Institute for Research and Applications of Fuzzy Modeling, Ostrava, Czech Republic

There are people who will live with us forever --- they influenced our life and stay next to us whenever a situation connected with them occurs. Our dear friend Da Ruan belongs to such people. We want to share with the reader memories of few meetings with him which we will never forget.

Irina's Memories

The first virtual meeting

In 1993 thanks to the Soros Foundation there appeared an opportunity for Russian scientists to attend foreign conferences. Being supported by this foundation, my colleague Alexander Ryjov participated at the Fifth IFSA World Congress in Seoul. After returning home he answered our questioning about everything: scientific level of the Congress, leading scientists, actual problems, etc. I do not remember all of his answers, but remember well only one of his remarks: ``a crazy and enthusiastic young Chinese guy invited me to become a member of the program committee of FLINS Workshop in 1994". At that moment, the idea standing behind FLINS (Fuzzy Logic and Intelligent Technologies in Nuclear Science) seemed unrealistic to us. None of us believed that a new science would be able to break into so conservative scientific area as nuclear science. But the "crazy Chinese guy" --- Da Ruan --- believed this idea. Thanks to his self-will, the Workshop came to the reality and took place in a beautiful Belgian town Mol in September 1994.

The first actual meeting in 1994

I attended that Workshop and remember its noble atmosphere. Two groups of well recognized scientists in nuclear science and fuzzy logic discussed possibility to apply new ideas to classical problems of automatic control. The third group of young enthusiasts (to which I belonged) proposed their ideas which should (according to our knowledge and strong belief) improve state of the art. Besides Da, I got acquainted with other members of the third group, and two of them --- Bernard De Baets and Gert De Cooman, became my good friends.

After the 1st FLINS nobody of us doubted that this conference would have a successful future. Da proved himself to be an excellent organizer, manager and researcher --- all-in-one. He was so charismatic that despite his young age people considered him as a leader. Of course, he understood very well his responsibility for all those whom he attracted for a joint work or scientific collaboration on various levels. He never promised what he was not able to realize and kept his word if he promised something. We can only imagine how hardly he worked in order to have this reputation.

The second meeting in 1996

Being motivated by the first successful FLINS I began to look for scientific contacts in Moscow academic spheres in the nuclear research area. We agreed with Da that our proposals will be included into a European project which he was going to submit. However, to create proposals in the area to which I did not belong was a big problem for me. Moreover, in the area of nuclear research people did not have any experience with fuzzy logic. Nevertheless, I agreed with a small group of researches that we would attend the 2nd FLINS workshop and formulate to Da our proposals.

Our trip to the 2nd FLINS workshop was more an adventure than a business trip. We did not find a support from Russian ministry and so, we had to finance everything from our pocket. Therefore, we took a tourist bus from Moscow to Brussels and travelled almost three days across Europe. We came to Mol (the conference site) at three o'clock in the morning, and despite this early time Da came to meet us.

I will never forget the warm acceptance which we had during and after the 2nd FLINS workshop. We were hosted in the guest house of the SCK CEN research center, visited its laboratories and were introduced to the running scientific programs. I remember our visit to the laboratory where Da worked together with his PhD students. They explained us details of their project and showed us various simulations. We appreciated very much the deep level of knowledge in both areas --- the fuzzy logic and the nuclear engineering, which Da and his students demonstrated to us.

We discussed our proposals which were not as well prepared as it was required by European committee. At that time we did not have enough experience to compete on the European level. Da was the youngest member of our scientific team which also included well recognized Russian specialists in the area of nuclear science. Despite of his age and high scientific level of my Russian colleagues, Da fairly explained that the proposal need to be prepared in more detail so that it is clear to everybody what and how would be achieved. I will never forget this lesson which helped me later in my scientific career.

We agreed that we start with a bilateral scientific cooperation and submit our joint proposals in the form of project to the Russian Ministry of Nuclear Energy. The project was approved on all levels, but finally was not financed due to the difficult economic situation at that time. We lost the great opportunity!

Another event during our stay in Mol happened at that time. Namely, Da and Greet announced their engagement! I still see their happy faces and how nicely they looked together. All of us wished them a happy family life. I have only one photo with them which remembers me that time.

B. Moiseev, Da and Greet in Mol 1996. P. Wang and Irina are at the back.

I attended several other FLINS conferences after those two in Mol. But this time, we participated in them together with my husband Vilém Novák.

Vilém's Memories

Meeting in 2002

I met Da Ruan first in 2002 at, as one immediately can guess, FLINS conference which took place in Ghent. I can only confirm everything what Irina wrote about his enthusiasm and charisma supported by the deep knowledge. We immediately became friends. I also met his wife and his family. It was nice to see happy people together.

At this occasion, we also had an opportunity to experience big influence and esteem which Da had in China. In Ghent, he introduced us to Prof. Xu Yang, well known mathematician from Southwest Jiaotong University in Chengdu. We had several discussions together about the, so called, fuzzy logic with evaluated syntax, to which Prof. Xu contributed independently on us. Under the influence of Da, Prof. Xu invited both Irina and me to Chengdu for one month of research stay. He prepared for us wonderful time full of exciting combination of work (we gave 12 lectures during our stay) and sightseeing in Chengdu and its surroundings. Thanks to Da, we enjoyed this unforgettable experience.

Meeting in 2007

In 2007, our institute organized 5th Conference of the European Society for Fuzzy Logic and Technology (EUSFLAT) in Ostrava. It was a big challenge for us though we had already experiences with organizing 7th IFSA World Congress in Prague in 1997. One of important problems to be solved was the choice of invited plenary speakers. Our idea was to combine theoreticians with applied scientists. The theoreticians were decided quickly. But applied scientist was a problem since we wanted somebody with sufficient charisma and experiences to attract the whole audience. But to be frank, we did not need to think long. We quickly realized that the best man who could fulfil our expectations was Da. An indeed, this was very good choice. Da arrived with full optimism and good mood and gave excellent talk followed by rich discussion. We enjoyed his presence and decided that we would invite him again at the earliest opportunity.

Meeting in 2009

Such an opportunity appeared two years later. In 2009, we organized a small International Workshop with the title ``Soft Computing: Where Theory Meets Applications". The workshop took place in a beautiful place in the mountains Beskydy, about 40 km from Ostrava. We invited only a small group of people whom we knew to be good theoreticians having at the same time also experiences with applications.

An indeed, this workshop turned out to be one of the most pleasant meetings which combined high scientific level with friendly atmosphere. Da was one of the invited speakers and his interesting talk was a strong source of inspiration for all of us. We had several hot discussions and also agreed on a closer cooperation.

D. Ruan, L. Stefanini, I. Perfilieva and V. Novák in Beskydy 2009

Unfortunately, Da left us too early. We will never forget this highly respectable nice man, good scientist, good friend, and loving husband and father.

4.15 Da Ruan and the World Council of Nuclear Workers

Professor Emeritus André Maïsseu

Président, World Council of Nuclear Workers, WONUC

On Monday, September 17th, 2007, the General Assembly of the International Agency for Atomic Energy (IAEA) was abuzz with rumours. The year 2006 had been marked by an exceptional event, the awarding of the Nobel Peace Prize to Mohamed El Baradei, the Director General. And everyone had his own thoughts, his own opinion, his comments, his exegesis, remarks and observations on the repercussions that the prize would have on the future of nuclear energy across the world, on the reactions of the so-called ecologists to the prize, and on the future of El Baradei when his term as the head of the IAEA came to an end ...

At the opening session, the score of members of the Board of Administration came along in single file, in a procession where monotony vied with their rhapsodic platitudes. Then came the usual ballet of the delegations, eager to present the nuclear programme of their countries, their proposals for a safer world, the right to nuclear energy for some of them, and its prohibition for others, the various definitions of the others, reversing according to whether the delegation at the podium was pro-Israel or Arab-speaking. Yet the same words systematically returned in a multiplicity and albeit constant identity of form, each member was keen to congratulate the new Nobel Peace laureate, in the most original, the most flattering, and the most eulogistic way possible, reflecting the many faces of hypocrisy.

And they all wondered whether the Royal Academy of Sciences of Sweden would be present to mark the great esteem of its members for the new laureate, although the vast majority of the delegates were unaware that the attribution of the Nobel Peace Prize was the domain of Norway and not of Sweden.

On the television screens scattered throughout the building where the conference was held, recalling the programme of the day, an announcement was quickly registered by all. It said that in the afternoon, during the plenary session, Mohamed El Baradei would receive the Atoms for Peace prize, from the name allocated to the speech given by US President Dwight Eisenhower to the United Nations on December 8th, 1953, and which led to the creation of the IAEA. The rumours grew, and along with them the idle talk and hearsay of those who always know more, distorting, revisiting the information, to restore it in a false-true reality fabricated from fictions and inventions. And since, in a world populated by diplomats, everyone is always obliged to know a little more than his neighbour, the word spread that in addition to awarding the Nobel Peace Prize, the Royal Academy of Sciences of Sweden was going to award to Mohamed El Baradei the famous and henceforth illustrious Atoms for Peace prize.

Barely a handful had remarked that this prize was to be awarded by the World Council of Nuclear Workers. And among these slightly more attentive readers, it was quickly added that the nuclear workers were cited simply because El Baradei wanted to associate the nuclear workers with the ceremony, so that the smallest contributors would not be overlooked at this great celebration, which consecrated nuclear energy as a peaceful form of energy.

Away from the hullabaloo, the members of the delegation of World Council of Nuclear Workers were discussing the precise organisation of the ceremony with the IAEA's protocol division, its legal department, the general secretariat, the private secretary of the director general, and above all, the security services. Who would take the floor, for how long, in what order etc. A consensus was rapidly reached on the name of the Japanese representative of WONUC, who would describe the framework surrounding the creation of this prize by WONUC, so as to mark the homage that we all owed to the Japanese people who had suffered personally from the horrors of the militaristic disasters of nuclear energy. They all hoped that the President of WONUC would then take the floor.

And suddenly, in the middle of the meeting, breathless, still holding his suitcase, barely landed from a plane flying from Mumbai or from Hong Kong, entered Da Ruan, China's representative of WONUC. He was absolutely determined to attend. In his view, as in the view of the other WONUC representatives, the creation of the Atoms for Peace prize by WONUC, and the choice of the IAEA's Director General as the first winner, was highly symbolic of the importance of nuclear energy in the search - beyond the illusory sustainable development bandied about to the point of indigestion by the ecologically-politically correct - for a lasting peace for humanity. Da Ruan had met André Maïsseu, the president of WONUC, for the first time at Santos in Brazil, at a conference where Dr Maïsseu had come to present a model of economic development in which the mathematical portion was based on fuzzy logic. During the ensuing period, Da Ruan and André Maïsseu, who became inseparable, explored all the beach restaurants of Santos, to enjoy the thousand and one ways in which the Brazilians, as worthy Lusitanophiles, processed cod, and to debate fuzzy logic, neuron networks, cognitivism and other connectionisms.

Moments later, it was universally agreed that Da Ruan should be the second speaker, and that the Atoms for Peace prize should be awarded to El Baradei in the form of a medal and a (small) cheque. Da Ruan energetically refused, arguing that this was a role of the president of WONUC and not his own. Only at the very last minute, bowing to the insistence of all, did Da Ruan agree.

His address to the 3000 delegates present at the General Assembly of the International Agency for Atomic Energy was full of subtlety, tact and intelligence. He gave his speech in English, with a few sentences in French and Mandarin.

It was then that the confusion that prevailed in many minds between the Nobel Peace Prize and the Atoms for Peace prize took an exciting turn. The ceremony was over, the dance of the aspirants to a Nobel Prize in Physics, Chemistry, Medicine and Economics reached its apogee. All the candidates who were seeking the ultimate recognition of their efforts, necessarily (in their opinion), the most complete, most grandiose, most miraculous that humanity had been able to generate, crowded around Da Ruan, inviting him to dinner, inviting him to come and visit them at their universities, all expenses paid – obviously – and to this and to that event. All these aspirants to supreme glory and immortality – and perhaps the accompanying cheque – were convinced that Da Ruan was THE representative of Sweden's Royal Academy of Sciences, with the ultimate power to award these Prizes.

Da Ruan was initially amused. Then, the waves of wily hypocrisy annoyed him. And we all finally got together, the entire delegation of WONUC, all hands aboard, to dine among true friends and to have a good laugh, that same evening, at a small inn at Grinzing, in the Vienna hills, where, according to the tourist brochure, Beethoven composed his 6th Symphony. As to the robe of the new white wine that we quaffed, it was necessarily fuzzy!

Da awarding the Atoms for Peace Prize to Nobel Prize winner El-Baradei at IAEA in Vienna on September 17, 2007

4.16 My Meetings with Da

Dr. Isis Truck

University Paris 8, Paris, France

The first time I met Da was through a student and colleague of mine in Blankenberge for FLINS 2004. They exchange about our paper on intelligent image retrieval through linguistic color expressions.

Then I met him in Madrid for FLINS 2008. I presented a paper with my PhD student on fuzzy logic coupled with service-oriented architectures. We talked about mixing fuzziness and GAI (Generalized additive independent) networks, such as the conditional preference networks.

In 2009, I went to Hasselt, Belgium for ISKE and we met again. He was one of the editors of the proceedings and we talked with a colleague of mine about the way to put into practice fuzzy concepts (we proposed there a fuzzy library dedicated to musicians and/or performers).

Last time I saw him was in Chengdu for FLINS 2010. I talked with him and with his wife Greet. We started to discuss projects together, contributions in journals, European project proposals (COST), etc. I will never forget these discussions and the talks we had.

4.17 Guiding Chinese Scientists to Germany

Prof. Dr. Wolfgang A. Halang

Fernuniversität in Hagen, Hagen, Germany

It must have been in September 1999, that I travelled to Zittau to attend a cosy little conference on fuzzy logic. Zittau is the small city in the most south-eastern corner of Germany neighbouring Poland and the Czech Republic. The conference was not the only reason to go there, but also that my father's family originates from a town nearby. On the conference's first day I went to my hotel's breakfast room. Having had already long-term relations to China, particularly Hong Kong, and with many Chinese people, I sat down on the table of a Chinese looking man. As expected, he was an attendant of the conference as well – it was Da Ruan.

We started to talk about our respective backgrounds, our relations to China and our research interests. Da told me that he went to China each summer, not only to visit his family, but also to give lectures and, most importantly, to search for candidates he could hire as postdoctoral fellows to work with him at the Belgian nuclear research centre. Postdocs – this approach raised my interest. Why bothering around with doctoral students who still need a great deal of support, especially when it comes to formulating their papers and theses? I thought it was a good idea to work with more mature and experienced people. In order to try it myself, I asked Da to keep in mind during his next scan for talents in China that I was also prepared to host a postdoc. And Da delivered sooner than expected.

In the summer of 2000, Da returned to his hometown Shanghai, and in the course of this he gave lectures at Shanghai Jiaotong University. One of his students there was Zhao Yi, who was to defend her PhD thesis just a few months later. Da identified her as a candidate for his personal "postdoc programme". Then, his plans happened to change. In 2001, he would leave Mol for a sabbatical with the nuclear research facility in the Norwegian city of Halden. Therefore, he suggested to me to look into the possibility of hosting Zhao Yi as a postdoc in Hagen.

Since I had made a bad experience with a Chinese student before, I decided to meet Zhao Yi in person before inviting her to Germany. As I was spending some time in Hong Kong around Chinese New Year 2001, we agreed to meet in Shenzhen on 1 February. It was a grey, hazy morning that I crossed the border to mainland China and walked the short distance to the Shangri-la Hotel. I had just sat down in the lobby and prepared to read the papers brought along, when a young couple stood up from a couch nearby and approached me. It was Zhao Yi with her husband Shi Jianliang, whom she had married just two weeks earlier, but of whose existence I did not know. We talked for two hours before we went up to the hotel's rotating restaurant to have lunch. Being the only guests there, all waiters came together to observe a very special intercultural exchange: I showed Shi Jianliang how to eat long noodles the Italian way with fork and spoon.

First, it appeared difficult to me having to host two persons now instead of one. But the solution came by itself. In April I received a letter from a research institute in Duisburg, not too far away from Hagen, which was looking for people with exactly Shi Jianliang's expertise. The rest was extremely easy: I called the letter's signee, Prof. Thomas Kaiser, he met the couple in Shanghai the same summer and they came to Germany in November 2001. After having lived for four years in Duisburg, they moved to Nuremberg where Dr. Shi took on a new job. The couple has now two daughters, Zhao Yi became a German and is still working with me.

Also still working with me since exactly the same time, and also put in contact with by Da Ruan is Prof. Li Zhong. Da's first FLINS conference in 1994 was attended by Prof. Ren Ping of Jinan University in Guangzhou, who recommended Li Zhong as his best MSc student at that time. In turn, Da invited Li Zhong to come to Mol as a visiting scientist. Although this visit never materialized due to first administrative difficulties and later Da's sabbatical in Halden, between 1996 and 2001 they co-operated rather intensively by correspondence. In early summer 2001, Da brought me in contact with Li Zhong, who had moved as a postdoc to Yonsei University in Seoul after having obtained his PhD from South China University of Technology in Guangzhou. Impressed by Li Zhong's scientific achievements and his extraordinary potential, I immediately strove to bring him to Hagen as a visiting scientist, although our research areas do not coincide. This could easily be achieved, because at that time we had a well-funded Sino-German exchange project. Later Li Zhong became an Alexander von Humboldt research fellow and a professor, as well as a father of a son and a German citizen.

The third Chinese person joining my group in autumn 2001 was Zhang Wei. Needless to say, that the contact with him came about via Da, and via one of Zhang Wei's colleagues at Southwest Jiaotong University in Chengdu who had been a visitor in Mol. After obtaining his PhD in early 2006, Zhang Wei stayed in Germany and found a job in industry. He also became a German citizen. With his wife and daughter he lives now in Hamburg, where he is responsible for issues of functional safety in a company manufacturing wind power systems.

My relation to this book's editor was spawned, in turn, by one of the contacts above Da had established: During a FLINS conference in Belgium, Li Zhong met Prof. Jie Lu and raised her interest in my chair. Her first visit in 2007, and now she is a Mercator Visiting Professor.

4.18 Da: My Best Friend and Saviour (贵人)

Prof. Dr. Zhong Li

Faculty of Mathematics and Computer Science, Fern Universität in Hagen, Germany

During the two weeks from October 29 to November 12, 2011 when Jie Lu came to Hagen, Germany to visit us, and we went to Mallorca, Spain for the Workshop Autonomous Systems 2011 and Prof. Wolfgang Halang's 60th birthday celebration, we could not stop talking about Da, that was really all about Da. Every time it was ended with a sigh. Actually, Da and his family were also invited to this event, and Da had already booked tickets. If Da were still alive, he could be with us chatting, discussing, and bringing us much fun just as before. We had even been longing for such a scene that Da could be with us on the island, but the reality has always dragged us into a deep sorrow. Sigh!

Since Da passed away on July 31, 2011, his friends, colleagues, and students (from famous professors to Ph.D. students) have been falling into deep sadness. I am wondering what a person Da was, who can attract so many people to commemorate him forever, including me.

My story regarding Da can be traced back to the year of 1994 when I was still a Master student at the Jinan University, Kanton under the supervision of Prof. Ping Ren. It must be someday in September 1994, Prof. Ren told me that he, Prof. Yingming Liu, and some other professors would go to Belgium for a conference (it must be FLINS, but the name did not come into my notice). I just remember that, as Prof. Ren came back, he told me that the conference organizer is a very smart and diligent guy, his name is Ruan Da.

As time flying away, the year shuttled to 1997, when I was just enrolled as a Ph.D. student at the South China University of Technology, Kanton and has got to know my girl friend (now my wife), Mei Juan, Prof. Ren asked me if I would like to pursue a Ph.D. degree with Da at the SCK·CEN. I still remember that I was so excited, because it was my dream to study overseas. But I could not accept this invitation due to some reason, which I planned to explain Da later in an appropriate occasion, but there will never be an opportunity for me. It seemed that my story about Da should stop here, because at that time Europe and Da seemed so far away from me, not only geographically but also psychologically.

After getting my Ph.D. degree, I went to Seoul, South Korea, on September 18, 2000, as a postdoctorate at the Yonsei University, accompanied by my wife, but with an ultimate target to the US. Very soon, in December 2000, I got an offer from the US; I was happy and also worried about that it would be too early to ask for resign. I decided to try to kill the clock, and until the end of March 2001, I had to take action to prepare to apply for visa. Man proposes, God disposes. The

China-U.S. aircraft collision incident on April 1, 2001 had messed up my plan. We tried twice to apply for visas early April and then in the middle of July, and the results were negative. I was at the end of my rope, nowhere to go. In the following weeks, I had been falling into deep desperation. Suddenly, Da came to my mind. I wrote him a short e-mail to ask him if he had any chance for me in Europe. Unexpectedly, he replied me immediately, and told me that there might be something for me. In a few days, Prof. Halang wrote to me that he could invite me to Hagen, Germany. Within a month, we had got everything done and I came to Hagen on September 18, 2001, precisely after one year's stay in Seoul. A new life has then begun. In Hagen, I got the famous Alexander von Humboldt Scholarship, got the title of Dr. of Science (Habilitation), got the German citizenship, and have become a Junior Professor and now an apl. Professor. It is just like that described in a poem from the Southern Song Dynasty – "After endless mountains and rivers that leave doubt whether there is a path out, suddenly one encounters the shade of a willow, bright flowers and a lovely village" (山穷水尽疑无路，柳岸花明又一村), and Da is that person, whom we name "Saviour" – "贵人" in Chinese –, for me.

It was followed with FLINS coming logically into my life. In September 2002, Prof. Halang sent a group of five staff members and students, including Zhang Wei, Zhao Yi, and me, to attend FLINS'02 in Ghent, Belgium, where I met Da for the first time, and was immediately impressed by his kindness, passion, humour, and perfect English. Although I am not directly working in the scope of FLINS, I told myself that I would adhere to participate in FLINS, not only due to the food of FLINS but also due to the amiable atmosphere at FLINS. FLINS'04 in Blankenberge, FLINS'06 in Genova, FLINS'08 in Madrid, and FLINS'10 in Chengdu, it is like a platform, on which I have got to know many best friends. Gradually, FLINS for me is not only an academic conference, but more like a family get-together. I remember clearly that last year on the E-Mei mountain, Sichuan, China, Da, Guangquan, and I sat together in the lobby of the Hong-Zhu-Shan Hotel, the conference site of FLINS'10, and chatted, I made a joke with Da and asked him to award me a medal at the FLINS'12 in Istanbul for my continuously following FLINS for ten years. Little does one think that Da took this seriously and noted down in his notepad. I talked this funny thing to other friends. Now, I do really hope that Da could award me a medal in Istanbul; but now where is Da? Shall I still go to Istanbul?

When I close my eyes, Da can still appear in front of me, life likely. Even in June 2011, Da has written a recommendation letter for my promotion to apl. Professorship. In my e-mail box, there is still a stack of Da's Emails about review assignments, project discussions, his working plans, etc.. Da had simply taken too much workload, now I think Da can relax himself finally, and let us share in.

Da, I wish you all the best in heaven with God.

Da and Zhong at FLINS2002 in Ghent, Belgium

4.19 Da Ruan in and with Aachen

Prof. Dr. Hans-Jürgen Zimmermann

Aachen Institute of Technology Operations Research, Aachen, Germany

When I met Da Ruan the first time in 1990, serving on his PhD committee, I was impressed by the quality and insight of his thesis and his argumentation. I did not expect, however, what a long and fruitful cooperation would follow. His PhD thesis focussed on a topic that was of basic importance to fuzzy inference and in which I was personally very interested. Very soon after this Da became an author of the leading international journal in the area of fuzzy sets, Fuzzy Sets and Systems, and very soon afterwards already an esteemed referee in Fuzzy Sets and Systems, and also for the European Journal for Operations Research, on whom I could always rely.

Already in 1995 he himself became the Guest Editor of a special issue of Fuzzy Sets and Systems on uncertainties and the application of fuzzy technology in nuclear engineering, which contained contributions of leading scientists in this area. This special issue was connected with the now-a-days well known stream of conferences called FLINS (Fuzzy Logic and Intelligent Technologies in Nuclear Science) the first of which Da organized in 1994 in Mol (Belgium). I had the honour and pleasure to be invited to this 1994 conference, which started an intensive and long lasting cooperation between Da Ruan in Mol and myself at the Aachen Institute of Technology. The proceedings of this first FLINS conference, the first of a series of prominent international conferences that took place in Mol and still takes place in several countries and continents was published –co-edited also by Ruan Da – as a book under the title " Fuzzy Logic and Intelligent Technologies in Nuclear Science", published by World Scientific in Singapore. Only two years later Da edited the book "Fuzzy Logic Foundations and Industrial Applications" in the International Series in Intelligent Technologies, published by Kluwer Academic Publishers, and kindly dedicated it to me. Many thanks again for this honour!

In parallel to these conferences in Mol, Da Ruan supported strongly the EUFIT conferences in Aachen by contributing to EUFIT 1993, 1994, 1995 and 1997. He not only supported these conferences, but, in spite of all his other obligations, he also edited a book in Chinese that contained some of my major publications and had the title "Fuzzy Sets and Operations Research for Decision Support". Obviously this book could never have been published without the active support of Da Ruan! Thanks again, Da!

Even though not directly related to the Aachen-Mol relationship I would also like to point to the facts that Da became a member of the editorial board of Fuzzy Sets and Systems, edited further special issues of this journal and co-authored additional papers in it.

I am, of course, aware of the many of the other things Da achieved in his life. Be it the many conferences that he contributed to, be it his many publications or the many universities that he was affiliated to. Compared to that, our common activities may seem minor. For me, however, they have become a very important part of my life.

Looking back to my first meeting with Da Ruan I cannot but realize how he developed from a young and promising PhD scholar to an internationally recognized scientist and practitioner, not only in the area of fuzzy sets but much beyond it. He also had become a very esteemed colleague and friend of mine and it leaves me with a deep feeling of sadness that he passed away that early, much too early! Thank you very much, Da, for all you have given me!

The jury of Da's PhD with E. Kerre, P. Smets and H. Zimmermann in 1990

4.20 Dear Da, We Miss You Forever

Prof. Dr. Ashok Deshpande

Founding Chair: Berkeley Initiative in Soft Computing (BISC)-Special Interest Group (SIG)-Environment Management Systems (EMS)
College of Engineering, Pune, India

Dr. Da Ruan is considered as one of the doyens in the area of Decision Research. Deep interest in soft computing based energy policy research brought us closer. We met at the World conference on Soft Computing in San Francisco where he presented an excellent paper in my session. I was also one of the panelists in his well organized panel discussion. I knew Da for eight years from the year 2004, when we started research collaboration together. Our collaborations were fruitful and in that time we co-authored a number of papers in intelligent decision support, decision and policy making under uncertainty, the applications of the Evidential Reasoning approach for nuclear waste management and cost estimation support for large nuclear engineering projects.

Da was so kind that in spite of his busy schedule, he promised me to be with us in November 2011 for the International Conference in India but that was not to happen. The organizers were excited and were expecting a mail from Da but my friend Jie Lu informed about the sad and untimely demise of our beloved friend Da.

Dear Da

You are with us. Now you must be busy in sharing experiences with those who are in heaven. I am confident that they will all appreciate your brilliance coupled with excellent human qualities. I go through your affectionate mails several times and cry almost all the times. You have won our hearts and will do so even in heaven.

4.21 My Memory of Da

Prof. Dr. Gabriella Pasi

Università degli Studi di Milano Bicocca, Italy

My last memory of Da brings to mind several images of San Francisco last May. We were both attending the WCSC conference, and I can see Jie, myself and Da seated close to each other on the bus going with the conference attendants to the conference banquet. Jie is on my left, Da is on my right, sometimes talking to us and sometimes looking at his PC…

I met Da several years ago, and although we never co-authored a paper together, several times we had the opportunity to exchange ideas and to collaborate on the definition of joint research proposals and research plans.

When I decided to start my journey in the beautiful land of research, I was very young and I had an "ideal" picture in mind about people going about in this land and about their interactions. I imagined a world (the world of ideas and intellectual exploration) where people could be free to focus on their main aim of conquering new knowledge and to go around inspired only by a great passion for research and noble objectives. What better world to interact and feel free in than in the world of thoughts, of ideas, of knowledge, of creativity, where there is room for everyone trying to make a contribution? Over the years and with experience I discovered that not all people in this environment are animated by the same objectives; wrong competition mechanisms can sometimes lead people to miss big opportunities for freedom and constructive cooperation. However, in a world that still remains beautiful for me my journey has brought, among others, an invaluable and precious gift: the opportunity to meet beautiful people, who have enriched my life. Da was one of these people: in my mind he is a man who truly represented the ideal prototype of a scientist: an open minded man, with intellectual honesty which allowed him to enjoy his experience in the world of research, while never forgetting the fundamental value of friendship, family and cooperation. He has made an invaluable and important contribution to the growth of his research field and to the people who worked with him.

For this reason he leaves an unforgettable trace of his passage in this life experience.

4.22 A Cherished Memory of Prof. Da Ruan

Prof. Dr. Peijun Guo

Faculty of Business Administration, Yokohama National University, Japan

I came to know Da fifteen years ago in a very beautiful city, Prague when we both attended the 7th International Fuzzy Systems Association World Congress (IFSA 1997). It was my great pleasure to meet some Chinese colleagues, especially Da, in Prague as it was still relatively rare to see Chinese academics in Europe back then. At our first meeting, Da impressed me with his sunny and pleasant smile. After chatting with him, I was even more impressed with his kindness and generosity. Da graciously offered me a ride for sightseeing around Prague as he drove to the conference from Belgium. After learning that I was a doctoral student of Professor Hideo Tanaka, he commented that they share a great colleague and good friend, Professor Zimmermann and we belong to an extended academic family. He went even further to promise assistance whenever such a need arises. From then on, I always feel and cherish his brotherly care and support.

In 1999, I sent Da an inquiry email about the possibility of conducting postdoctoral research in Europe in case that I could not find a suitable position upon my graduation next year. He immediately responded with an affirmative answer. He was truly a gentleman who was always ready for helping others. This proposed trip did not materialize as I successfully secured an academic position at Kagawa University upon obtaining my Ph.D. degree in 2000. Nevertheless, I still greatly appreciate Da for his generous offer.

In 2001, I attended the 2001 European Operational Research Conference in Rotterdam. I brought him a bottle of Japanese liquor as a gift. He was very happy and mentioned it at different occasions for several years. Da was such a considerate man who always remembered what you had done for him, no matter how negligible it was.

In 2004, I experienced a nightmare while attending the 6th International FLINS Conference in Blankenberge: my suitcase was robbed around the Bruxelles-Central station. I lost virtually everything such as my passport, credit card, and wallet. Da gave me numerous invaluable help, and his remarkable care and kindness alleviated the torment caused by that terrible incident.

Recently, I had the distinguished pleasure to meet Da in several International Conferences on Intelligent Systems and Knowledge Engineering held in Chengdu (ISKE2007), Xiamen (ISKE2008) and Hangzhou (ISKE2010). We greatly enjoyed each one of these gatherings. We usually held parties with other Chinese friends and Da was always an incredible mood-maker. His enthusiasm, sincerity, wisdom and humor always created a perfect atmosphere for everyone to enjoy the party. At these pleasant gatherings, we shared academic ideas as well as life feeling and experience. Such talks often continued beyond midnight and I have learned tremendously from him. Everything imprints in my mind and still seems to me as happenings yesterday.

The time I spent with Da was always blissful. I am deeply saddened for losing a most respectful friend and a caring brother. I cannot help shedding tearing in thinking of you. Da, I greatly cherish those sweet memories of you and will miss you forever. May you rest in peace.

Prof. Da Ruan with Peijun Guo at the 2008 ISKE Conference, Xiamen, China

4.23 Da Ruan: A Great Scientist, a Scholar, and Much More

Prof. Dr. Janusz Kacprzyk

Full Member, Polish Academy of Sciences
Foreign Member, Spanish Royal Academy of Economic and Financial Sciences
Fellow of IEEE and IFSA,
Systems Research Institute, Polish Academy of Sciences, Ul. Newelska 6, Warsaw, Poland

It is difficult to write something non-trivial, beyond words of condolence and expressions of sorrow, when a person passes away with whom we have been close for years in many respects. Da Ruan, has been a close friend to me, indeed, and we have had a friendly professional and personal relation, with a mutual support and cross-fertilization.

He was younger, and our life crossed some 20 or more years ago. For years I have had close contacts with Ghent University, visiting quite often various teams, notably that of Professor Etienne E. Kerre. When I first met Da, who at that time was a Ph.D. student of Kerre, I was expecting a situation quite characteristic for virtually all Kerre's Ph.D. students, that is, a high level of mathematics but at the same time a rare ability to enter the field of advanced applications of advanced mathematical tools and techniques. This is a great ability of Kerre, not shared by all mathematicians, but so relevant. The attitude and ability of Kerre is similar to that of Stanislaw Ulam, a famous Polish and then American mathematician, who had both an illustrious career in theoretical mathematics and applied mathematics being one of fathers of the hydrogen bomb, the Monte-Carlo method, bioinformatics, etc. Many famous statements are attributed to him like: „The infinite we shall do right away. The finite may take a little longer" which was shocking in the years just before or after World War II when infinity was all what mathematics was about but which is crucial now, notably in view of an ever increasing role of computing in both applied and even theoretical mathematics.

Those, maybe a little bit long remarks are relevant to understand the environment in which Da was formed as researcher. This experience has had a tremendous impact on his professional career, as it is always the case of all young people who enter the academic career. This was the time when I met him for the first time.

Just after the successful defense of his Ph.D., Da Ruan started his professional career, first as a "post-doc" and then as the principal investigator at the Belgian Nuclear Research Centre (SCK•CEN). To many people, notably in Europe, this might have been an unusual career for a very good young Ph.D. because such people usually tend to follow an academic career at a university. To me, however, this not proper and I would recall here again Ulam, and also many other great mathematicians like John von Neumann, Richard Bellman, Allan Turing, etc. who have been associated both with top universities and famous research labs solving real world problems. Those people have combined a high level theoretical work with relevant, non-trivial applications. We have all profited from their work, sometimes in a negative way, as it contributed to the advances in military application, but mainly in a positive way exemplified by the advances in computing and ICT.

Da Ruan has found at the SCK•CEN a proper place to grow and has greatly contributed to the strengthening of its stature as a leading research laboratory. His major research interests were in the fields of mathematical modeling, theory and above all applications of computational intelligence methods, uncertainty analysis, decision support systems, safety and security related analyses, etc. As it is generally the case when a high proficiency in mathematics and computer science is combined with interesting and challenging real world problems, Da's works have quickly become examples of top quality landmark contributions. He was one of a few who knew how to integrate high level theory and real applications. This is, in my opinion, something extremely relevant for Europe, namely that there are people who "go the American way", i.e. do research at the highest level at a commercial, industrial or governmental research laboratory, not only at a university. There are more and more people like this in Europe, and Da was certainly one of the most prominent representatives of them. To me, a control engineer by training, and by necessity a person for whom applications, but not trivial ones! matter is extremely relevant.

Having quickly become a prominent member of the world scientific community, Da – with his unstoppable energy and vision – was constantly starting all kinds of new initiatives: new conferences, new journals, new working and interest groups, etc. In recent years it was practically impossible to find a conference or another scientific gathering, round table discussion, edited volume project, etc. in our community in which he was not present. And people were listening to him because he had something to say: relevant, full of deep insight and vision.

Over the years he was a frequent visitor to his homeland, Mainland China, giving lectures, bringing top Chinese graduate and Ph.D. students to Europe, and promoting all kinds of scientific exchange. In the Chinese scientific society he was greatly respected in spite of his young age. He was considered part of the elitist group of top Chinese born scientists working abroad, together with Professor Jie Lu, a coeditor of this important volume.

I could write much more about Da Ruan's brilliant achievements, scientific contributions, organizational skills, etc. However, this all becomes irrelevant when it concerns a friend because at the human and personal level we do not lose a scientist, scholar, researcher, expert, etc. but we lose a friend, as simple as it is. In my case this is exactly what has happened. I have lost a real and loyal friend, and a great human being. What can I say more? One can ask numerous questions like: why him, why so early, etc. This is however the mystery of life and death, beyond the human comprehension. Putting it shortly, we will all miss Da a lot.

Janusz and Da in FLINS2008 in Madrid

4.24 My Life Experience with Da Ruan

Prof. Dr. Rita Ribeiro

Department of Electrical Engineering, Universidade Nova de Lisboa, Portugal

I first heard about Da Ruan and his work on nuclear plants from a colleague that visited his lab around 1996. In 1998 I met him personally at a WAC conference in Alaska. Since then we met relatively regularly at conferences from IPMU, FLINS, EUROFUSE etc.

During the years from 1988 to 2008 all our encounters were quite interesting and I enjoyed his dry sense of humor! Moreover, because of his research interest in fuzzy multi-criteria we had quite interesting conversations in the corridors of the fuzzy conferences we both attended.

In 2009 we attended the Workshop on "Soft Computing: Where Theory Meets Applications", held in Trojanovice, Czech Republic where we had a lot of fun and we started collaborating more closely, which led to publishing a paper in the series where he was Editor "Computational Intelligent Systems" by World Scientific in 2010.

It was also during the 2009 workshop, organized by Irina Perfilieva and Vilem Novak that he shared a funny story about himself that made me laugh copiously. His real name was Ruan Da (Ruanda...). At the time of the massacres in Ruanda and the respective media exposure in the television, he suffered many humiliations in airports and other places, usually by being singled out and taken to a "private room for talking". ... Finally, he decided to change the name order (surname + name) and to his surprise he found out it was a wonderful decision: there are NO other Chinese person called Ruan (as a surname) and there are millions of Da(s)!! He became the only Chinese with surname "Ruan" and this also meant to be singled out when a search is made in the internet (Google it!).

The last time I saw him it was during the World Conference on Soft Computing 2011 in S. Francisco (USA), where he organized an interesting panel session about "challenges and future directions of Soft Computing", to which I contributed (by his request) but unfortunately missed. During this event he also told me that he was planning other activities (e.g. COST action) and I was really looking forward to further our collaboration.

Da Ruan: you will be missed by what has been and what could have been.

4.25 Da Was Near with Me in the Very Important Stages of My Life

Prof. Dr. Alexander Ryjov

Moscow State University, Moscow, Russia

The story of my acquaintanceship with Ruan Da started in April, 1993. At that time I met with Professor Etienne Kerre during the Second International Symposium on Uncertainty Modeling and Analysis, University of Maryland, USA. We easily found common ground in professional and personal levels. At that time in Russia there were a lot of intense and strenuous changes which opened new possibilities and horizons for cooperation with colleagues around the world. We decided to start a small research project.

After several months Professor Gert de Cooman from Ghent University, former Ph.D student of Professor Kerre, visited Moscow State University n.a. Lomonosov. We had an unforgettable time during several visits of professor de Cooman in the framework of the project; Gert became a close friend for me and my family. During one of the visits I got a proposal from Da Ruan to join the program committee of FLINS'94 – the first international workshop on fuzzy logic and intelligent technologies in nuclear science, planned in September 1994 in Mol, Belgium. I asked Gert to tell details about the workshop and Da Ruan. Gert's reply was very short: "If Da is doing something, it will be of high quality and great result." It was first words about Da I heard.

We decided to try to involve Russian scientists in the workshop. Our motivation was very simple.

It could be argued that Russian scientists more than others have been confronted with the necessity of managing nuclear objects in extreme situations, and controlling the non-proliferation of military nuclear technology. This is a consequence of the Chernobyl accident and the political and economic difficulties accompanying the disintegration of the USSR. As these management, monitoring and control problems are characterized by large uncertainty and are based on evaluations by experts providing diverse and inconsistent information, fuzzy and other intelligent technologies seem, in our opinion, to provide adequate methods for their solution. These problems, certainly, are of worldwide importance. Solving them successfully can save many lives and at the same time contribute to the improvement of the world's ecological situation.

These arguments turned out true. A short time after the call for papers for the FLINS'94 workshop, many Russian researchers submitted their proposals containing current work, and finished results within the scope of FLINS. Due to the scarcity and poor quality of the communication channels between Belgium and Russia, and probably also due to our lack of experience, at that time, in optimally exploiting what little means existed, the three of us had to put up with a fair amount of difficulties and hard work in gathering, proofreading and typesetting all the interesting material, and in making the accepted papers for FLINS'94 reach Belgium at all.

As we already indicated, the idea of using fuzzy systems and other intelligent technologies in the nuclear domain stirred up a large interest in Russia. Scientists there prepared twenty reports that were accepted by the organizing committee for presentation during the FLINS'94 workshop. This is more than thirty-five percent of all the accepted papers, which serves as an illustration of the Russian interest in FLINS. During the workshop, in Mol, Da introduced me to Professor Y. Nishiwaki and Mr. A. Fattah from department of Safeguards, International Atomic Energy Agency. First discussions which took place with participation Da, were the starting point for project entitled DISNA – Development of an Intelligent System for Monitoring and Evaluation of Peaceful Nuclear Activities. This project is the most valuable research project for me. The project was finished in 1999. During the system's development we faced the problem of choosing and tuning aggregation's operators in fuzzy hierarchical systems. After a series of feeble attempts, we decided to enlist someone who is capable to provide solution very fast. Professor Lotfi Zadeh, senior scientific advisor of the project, recommended me to ask Da Ruan. I was surprised how quick and how brilliantly Da and his team provided us the necessary solution. Thus Da made a valuable contribution to this important project.

Back to Russia in 2000, I kept my position in Moscow State University and joined Cadence Design Systems, Inc. – world leader in Electronic Design Automation tools. I was deputy general manager of Cadence Russia office, responsible for advanced research and education programs until 2007. At that time I was far away from fuzzy logic theory and applications; I have met with Da several times on conferences only. I was happy to see how Da became more and more mature scientist with very interesting projects and results; he deservedly received recognition as a world leading researcher in fuzzy logic theory and applications.

My last contact with Da was in 2009. At that time I was deputy director of Lebedev Institute of Precision Mechanics and Computer Engineering, Russian Academy of Sciences. Friend of me, professor Sergey Ulianov, came back to Russia after more than 10 years of working in universities and companies of Japan and Italy. We decided to start project with FW7 – the Seventh Framework Programme for Research and Technological Development. According to the rules, we needed partners from EU. I asked Da to help us with partner's network. Just after a couple of weeks from my first e-mail to Da, we had a confirmation from universities from Belgium, Italy, Germany, Portugal, Spain, Hungary, and UK. It was real visibility of authority Da had in research society.

Unfortunately, I cannot say that Da was my close friend. But Da was near with me in the very important stages of my life. Every time he was "on-line". A lot of my friends and colleagues in Russia remember Da and mourn the loss of Da.

4.26 Our Collaboration with Da

Prof. Dr. Francisco Herrera and Prof. Dr. Enrique Herrera-Viedma

Dept. of Computer Science and A.I, University of Granada, Granada, Spain

Da and the Spanish team at FLINS2008, Madrid

We were together with the Adjunct Professor Da in the 8th Int. FLINS Conference on Computational Intelligence in Decision and Control celebrated in Madrid, Spain September 21-24, 2008. In 2005 we invited him to give the talk titled "Soft Computing and Intelligent Systems in Industry: Implementation, Challenges and Future Directions" in the Seminar on New Trends on Intelligent Systems and Soft Computing in our Master in Granada University. We have co-edited a special issue with him in Information Sciences and also collaborated in the development of his journal Int. J. of Computational Intelligent Systems from the editorial board. He was a great researcher in our fuzzy community and a very good friend. He will always be in our memory.

4.27 Condolence on Adjunct Professor Da Ruan

Prof. Dr. Tijen Ertay

Faculty of Management, Istanbul Technical University, Istanbul, Turkey

Unfortunately, as his colleagues, we are distressful because of the fact that Adjunct Professor Da Ruan passed away on 31 July, 2011 from heart failure. I came together in FLINS organizations twice in past time. I have collaborated on some publications with him in past time. The last meeting has been FLINS'2010 in China. FLINS Conferences have identified with his name. This photograph belongs to FLINS'2010 Organization and last memory with his colleagues. This untimely death is a big missing for all science community and has shocked the colleagues in his research community and family. I believe that this book realized for Da Ruan' memory is to be reunited with his spirit and he sees us from a place that we could not see. In this like situation, the words are not enough to express sadness clearly. He always bears in minds with his excellent personality and the contributions to the fuzzy logic research fields.

Prof. Da Ruan and the Turkish team at FLINS2010 in Chengdu, China

4.28 In Fond Memory of Da Ruan

Prof. Dr. Dong-Ling Xu

Manchester Business School, The University of Manchester, UK

I knew Da for eight years from the year 2004, when we started research collaboration together. Our collaborations were fruitful and in that time; we co-authored a number of papers in intelligent decision support, decision and policy making under uncertainty, the applications of the Evidential Reasoning approach for nuclear waste management and cost estimation support for large nuclear engineering projects.

Prof. Da Ruan and Dong-Ling Xu at FLINS2006 in Genova, Italy

Da visited Manchester on several occasions and I also paid return visits to him at SCK·CEN a number of times. Over the years we also met in conferences and workshops in various parts of the world. Last year, my colleague Prof Jian-Bo Yang and I spent a month at SCK·CEN with Da and identified many research areas and topics of focus in what we had expected to be a long term collaboration.

To me and countless others, Da was not only an energetic and outstanding researcher, but also a warm-hearted friend. I can still remember during our discussions at SCK·CEN, his voice was so full of energy that one could almost feel the whole room was vibrating! He had so many highly productive collaborations and influential contacts that it seemed he had 72 hours to his day. He was so active, constantly organising conferences, book editing, journal publishing, or research proposals, that we would hear, read about, think of, or talk of his name almost every day.

Da worked hard, but he also enjoyed life to the full. He was an excellent bridge player, and bridge games on Friday evenings were a favourite pastime of his when he was not travelling. Last year while I was visiting him, there was a special event day at SCK·CEN in which Da participated in a poker game. Using just a tiny portion of his vast knowledge in statistics, he won the game quite spectacularly. He loved sea food very much and was an expert in enjoying Chinese tea. He is survived by his two lovely sons, of whom he was very proud, and a wonderful and caring wife, Greet. She and I shared an enjoyable friendship and many interests, and my heart goes out to her and the children. Da, we miss you dearly. May you rest in peace.

4.29 Da Always Ready to Help

Prof. Dr. Paul P. Wang

Duke University, Durham, USA

The last time I saw Da Ruan in Brussels, Belgium attending a Fuzzy Conference several years ago. While we were focusing in an interesting conversation about the conference organizing activities of the Fuzzy Business during its infancy and formation period, Da was interrupted by an emergency that three graduate students were mugged by some petty thieves who have had robbed the Chinese students using soft weapon of shaving creams. Not only the students no longer had their passports under their procession, all their cash for the trip were completely gone as soon as they walked out the Brussels train station. This incident immediately promoted Da into a crisis management urgent care mode. After several hours of back and forth between the police, the Embassy, and the students, apparently first time making an oversea trip, was finally settled to attend the International Fuzzy Conference.

Very few people understood and had the appreciation what Da Ruan has done than me. We both understood what the tasks demand you to be patient in problems solving, in addition to readily apologize when the things go bust! I encouraged him by saying, any movement or waging a battle with the mainstream must have the people like Saul, or Paul, should you choose and dare to be different! It took at least half century for Fuzzy Logic to become a part of main stream!

The time to say good bye finally came, I remember vividly that I offered Da for a rain check to stay with my family at our North Topsail Beach condo at St. Regis resorts for a few days to rest and to recover. No one, in my mind, deserves more than Da to have a break. Encouraged by noble goal, peer pressures, desire to be the very best, Da was simply carried away, and even not consciously aware what has been going on?

My heart felt sad!

Prof. Paul Wang, Da and other friends in JCIS conference, USA in 2005

4.30 In Memory of Da Ruan

Prof. Dr. Ronald R. Yager

Machine Intelligence Institute, Iona College, New Rochelle, NY, USA

It is with a great sense of loss that I am writing this small memoriam about Da Ruan. I believe Da, who was full of ideas and energy for new ventures, had a bright future from which all of us would have benefited.

The first time that I met Da was at a conference in Europe. While I can't recall the conference or city, I can vividly remember our encounter. Da approached me with a question about an issue he was struggling with at the nuclear agency. They were trying to evaluate the safety at some nuclear plants around the world, and they had to deal with the issue of missing information. In the course of our conversation I was impressed by Da's sophisticated evaluation of the problem as well as his deep concern for the problem of nuclear safety. Over the years, Da had discussed with me many problems which he encountered in his role at the nuclear agency. I always found the problems very challenging and stimulating. On a number of occasions they provided an impetus for some of my research investigations.

As I came to know Da, I slowly came to be aware of his many professional activities and personal interests. Over the years Da and myself developed a relationship that was both of a professional and friendship nature. Da was kind enough to invite me to a number of conferences he organized both in Europe and China. I was always pleased to attend these events as they were always of the highest quality. One impressive feature of Da was his ability to easily function effectively and comfortably in both the East and West.

In addition to our professional relationship, Da and I had a special connection as a result of our enjoyment of food, particularly Chinese food. As much as Da enjoyed life in Belgium, I think that the one thing he really missed was good Chinese food. Every time I was at a conference with Da in China, he always had some interesting restaurant for us to visit. One of my most memorable experiences with Da was a meal, although not a Chinese one, overlooking the ocean in Sydney. We went with the daughter of Jie Lu and enjoyed ourselves in this very trendy environment. I also remember fondly the Banquet at a Chinese restaurant at Jie Lu's conference in Sydney in which Da came around to our table and asked for the head of fish we had just eaten, as he felt this was a very special delicacy.

My last meeting with Da was at the World Conference on Soft Computing in San Francisco. Da with his usual energy organized a very stimulating round table on applications. In addition, Da was very pleased at the location of the conference hotel as it was next to San Francisco's famed Chinatown. This location enabled Da to have Dim Sum lunch which he enjoyed so much. My last memory of Da was his seating with Jie Lu at the banquet dinner, a dinner which like Da, bridged the gap between Europe and China. I will forever remember Da in this way.

In the course of writing this memoriam, I had the opportunity to discuss Da with my wife who also knew Da. Her comments focused on her impression of him as a kind and gentle person who was very considerate of other people. She also feels a great loss about not being able to meet Da at some future conference.

Da, Ron and Guangquan in AI conference, Sydney, in Dec. 2005

MIX
Papier aus verantwortungsvollen Quellen
Paper from responsible sources
FSC® C105338

If you have any concerns about our products,
you can contact us on
ProductSafety@springernature.com

In case Publisher is established outside the EU,
the EU authorized representative is:
Springer Nature Customer Service Center GmbH
Europaplatz 3, 69115 Heidelberg, Germany

Printed by Libri Plureos GmbH
in Hamburg, Germany